Everyone's Guide to Planet Neptune

Compiled by
Kiera Mcune

Scribbles

Year of Publication 2018

ISBN : 9789352979561

Book Published by
Scribbles

(An Imprint of Alpha Editions)

email - alphaedis@gmail.com

Produced by: PediaPress GmbH
Limburg an der Lahn
Germany
http://pediapress.com/

The content within this book was generated collaboratively by volunteers. Please be advised that nothing found here has necessarily been reviewed by people with the expertise required to provide you with complete, accurate or reliable information. Some information in this book may be misleading or simply wrong. Alpha Editions and PediaPress does not guarantee the validity of the information found here. If you need specific advice (for example, medical, legal, financial, or risk management) please seek a professional who is licensed or knowledgeable in that area.

Sources, licenses and contributors of the articles and images are listed in the section entitled "References". Parts of the books may be licensed under the GNU Free Documentation License. A copy of this license is included in the section entitled "GNU Free Documentation License"

The views and characters expressed in the book are those of the contributors and his/her imagination and do not represent the views of the Publisher.

Contents

Articles 1

Introduction 1
    Neptune 1

Discovery 29
    Discovery of Neptune 29

Physical characteristics 43
    Extraterrestrial diamonds 43

Orbital resonances 49
    Kuiper belt 49
    Neptune trojan 76

Formation and migration 83
    Formation and evolution of the Solar System 83
    Nice model 107

Moons 117
    Moons of Neptune 117

Planetary rings 131
    Rings of Neptune 131

Exploration **143**

  Exploration of Neptune 143

Appendix **151**

  References 151

  Article Sources and Contributors 157

  Image Sources, Licenses and Contributors 159

Article Licenses **161**

Index **163**

Introduction

Neptune

<indicator name="pp-default"> 🔒 </indicator>

Neptune

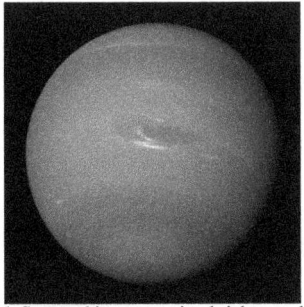

Neptune's Great Dark Spot and its companion bright smudge; on the west limb the fast moving bright feature called Scooter and the little dark spot are visible.

Discovery	
Discovered by	• Johann Galle • Urbain Le Verrier
Discovery date	23 September 1846
Designations	
Pronunciation	/ˈnɛptjuːn/ (🔊 listen)
Adjectives	Neptunian
Orbital characteristics[1,2]	
Epoch J2000	
Aphelion	30.33 AU (4.54 billion km)

Perihelion	29.81 AU (4.46 billion km)
Semi-major axis	30.11 AU (4.50 billion km)
Eccentricity	0.009456
Orbital period	• 164.8 yr • 60,182 days • 89,666 Neptunian solar days
Synodic period	367.49 days
Average orbital speed	5.43 km/s
Mean anomaly	256.228°
Inclination	1.767975° to ecliptic 6.43° to Sun's equator 0.72° to invariable plane[3]
Longitude of ascending node	131.784°
Argument of perihelion	276.336°
Known satellites	14
Physical characteristics	
Mean radius	24,622±19 km
Equatorial radius	24,764±15 km[4] 3.883 Earths
Polar radius	24,341±30 km 3.829 Earths
Flattening	0.0171±0.0013
Surface area	7.6183×10^9 km^2 14.98 Earths
Volume	6.254×10^{13} km^3 57.74 Earths
Mass	1.0243×10^{26} kg 17.147 Earths 5.15×10^{-5} Suns
Mean density	1.638 g/cm^3[5]
Surface gravity	11.15 m/s^2 1.14 g
Moment of inertia factor	0.23 (estimate)
Escape velocity	23.5 km/s
Sidereal rotation period	0.6713 day 16 h 6 min 36 s
Equatorial rotation velocity	2.68 km/s (9,650 km/h)
Axial tilt	28.32° (to orbit)

North pole right ascension	19h 57m 20s 299.3°
North pole declination	42.950°
Albedo	0.290 (bond) 0.41 (geom.)

	Surface temp.	min	mean	max
	1 bar level		72 K (−201 °C)	
	0.1 bar (10 kPa)		55 K (−218 °C)	

Apparent magnitude	8.02 to 7.78
Angular diameter	2.2–2.4"
Atmosphere	
Scale height	19.7±0.6 km
Composition by volume	• **Gases:** • 80%±3.2% hydrogen (H_2) • 19%±3.2% helium (He) • 1.5%±0.5% methane (CH_4) • ~0.019% hydrogen deuteride (HD) • ~0.00015% ethane (C_2H_6) • **Ices:** • ammonia (NH_3) • water (H_2O) • ammonium hydrosulfide (NH_4SH) • methane ice (?) ($CH_4 \cdot 5.75H_2O$)

Neptune is the eighth and farthest known planet from the Sun in the Solar System. In the Solar System, it is the fourth-largest planet by diameter, the third-most-massive planet, and the densest giant planet. Neptune is 17 times the mass of Earth and is slightly more massive than its near-twin Uranus, which is 15 times the mass of Earth and slightly larger than Neptune.[6] Neptune orbits the Sun once every 164.8 years at an average distance of 30.1 AU (4.5 billion km). It is named after the Roman god of the sea and has the astronomical symbol ♆, a stylised version of the god Neptune's trident.

Neptune is not visible to the unaided eye and is the only planet in the Solar System found by mathematical prediction rather than by empirical observation. Unexpected changes in the orbit of Uranus led Alexis Bouvard to deduce that its orbit was subject to gravitational perturbation by an unknown planet.

Neptune was subsequently observed with a telescope on 23 September 1846 by Johann Galle within a degree of the position predicted by Urbain Le Verrier. Its largest moon, Triton, was discovered shortly thereafter, though none of the planet's remaining known 13 moons were located telescopically until the 20th century. The planet's distance from Earth gives it a very small apparent size, making it challenging to study with Earth-based telescopes. Neptune was visited by *Voyager 2*, when it flew by the planet on 25 August 1989. The advent of the *Hubble Space Telescope* and large ground-based telescopes with adaptive optics has recently allowed for additional detailed observations from afar.

Like Jupiter and Saturn, Neptune's atmosphere is composed primarily of hydrogen and helium, along with traces of hydrocarbons and possibly nitrogen, but it contains a higher proportion of "ices" such as water, ammonia, and methane. However, its interior, like that of Uranus, is primarily composed of ices and rock, which is why Uranus and Neptune are normally considered "ice giants" to emphasise this distinction. Traces of methane in the outermost regions in part account for the planet's blue appearance.

In contrast to the hazy, relatively featureless atmosphere of Uranus, Neptune's atmosphere has active and visible weather patterns. For example, at the time of the *Voyager 2* flyby in 1989, the planet's southern hemisphere had a Great Dark Spot comparable to the Great Red Spot on Jupiter. These weather patterns are driven by the strongest sustained winds of any planet in the Solar System, with recorded wind speeds as high as 2,100 km/h (580 m/s; 1,300 mph). Because of its great distance from the Sun, Neptune's outer atmosphere is one of the coldest places in the Solar System, with temperatures at its cloud tops approaching 55 K (–218 °C; –361 °F). Temperatures at the planet's centre are approximately 5,400 K (5,100 °C; 9,300 °F). Neptune has a faint and fragmented ring system (labelled "arcs"), which was discovered in 1982, then later confirmed by *Voyager 2*.

History

Discovery

Some of the earliest recorded observations ever made through a telescope, Galileo's drawings on 28 December 1612 and 27 January 1613 contain plotted points that match up with what is now known to be the position of Neptune. On both occasions, Galileo seems to have mistaken Neptune for a fixed star when it appeared close—in conjunction—to Jupiter in the night sky; hence, he is not credited with Neptune's discovery. At his first observation in December 1612, Neptune was almost stationary in the sky because it had just turned retrograde that day. This apparent backward motion is created when Earth's

Figure 1: *Galileo Galilei*

orbit takes it past an outer planet. Because Neptune was only beginning its yearly retrograde cycle, the motion of the planet was far too slight to be detected with Galileo's small telescope. In July 2009, University of Melbourne physicist David Jamieson announced new evidence suggesting that Galileo was at least aware that the "star" he had observed had moved relative to the fixed stars.

In 1821, Alexis Bouvard published astronomical tables of the orbit of Neptune's neighbour Uranus. Subsequent observations revealed substantial deviations from the tables, leading Bouvard to hypothesise that an unknown body was perturbing the orbit through gravitational interaction. In 1843, John Couch Adams began work on the orbit of Uranus using the data he had. Via Cambridge Observatory director James Challis, he requested extra data from Sir George Airy, the Astronomer Royal, who supplied it in February 1844. Adams continued to work in 1845–46 and produced several different estimates of a new planet.

In 1845–46, Urbain Le Verrier, independently of Adams, developed his own calculations but aroused no enthusiasm in his compatriots. In June 1846, upon seeing Le Verrier's first published estimate of the planet's longitude and its similarity to Adams's estimate, Airy persuaded Challis to search for the planet. Challis vainly scoured the sky throughout August and September.

Figure 2: *Urbain Le Verrier*

Meanwhile, Le Verrier by letter urged Berlin Observatory astronomer Johann Gottfried Galle to search with the observatory's refractor. Heinrich d'Arrest, a student at the observatory, suggested to Galle that they could compare a recently drawn chart of the sky in the region of Le Verrier's predicted location with the current sky to seek the displacement characteristic of a planet, as opposed to a fixed star. On the evening of 23 September 1846, the day Galle received the letter, he discovered Neptune within 1° of where Le Verrier had predicted it to be, about 12° from Adams' prediction. Challis later realised that he had observed the planet twice, on 4 and 12 August, but did not recognise it as a planet because he lacked an up-to-date star map and was distracted by his concurrent work on comet observations.

In the wake of the discovery, there was much nationalistic rivalry between the French and the British over who deserved credit for the discovery. Eventually, an international consensus emerged that both Le Verrier and Adams jointly deserved credit. Since 1966, Dennis Rawlins has questioned the credibility of Adams's claim to co-discovery, and the issue was re-evaluated by historians with the return in 1998 of the "Neptune papers" (historical documents) to the Royal Observatory, Greenwich. After reviewing the documents, they suggest that "Adams does not deserve equal credit with Le Verrier for the discovery of Neptune. That credit belongs only to the person who succeeded both in predicting the planet's place and in convincing astronomers to search for it."

Naming

Shortly after its discovery, Neptune was referred to simply as "the planet exterior to Uranus" or as "Le Verrier's planet". The first suggestion for a name came from Galle, who proposed the name *Janus*. In England, Challis put forward the name *Oceanus*.[7]

Claiming the right to name his discovery, Le Verrier quickly proposed the name *Neptune* for this new planet, though falsely stating that this had been officially approved by the French Bureau des Longitudes. In October, he sought to name the planet *Le Verrier*, after himself, and he had loyal support in this from the observatory director, François Arago. This suggestion met with stiff resistance outside France. French almanacs quickly reintroduced the name *Herschel* for Uranus, after that planet's discoverer Sir William Herschel, and *Leverrier* for the new planet.

Struve came out in favour of the name *Neptune* on 29 December 1846, to the Saint Petersburg Academy of Sciences. Soon, *Neptune* became the internationally accepted name. In Roman mythology, Neptune was the god of the sea, identified with the Greek Poseidon. The demand for a mythological name seemed to be in keeping with the nomenclature of the other planets, all of which, except for Earth, were named for deities in Greek and Roman mythology.

Most languages today, even in countries that have no direct link to Greco-Roman culture, use some variant of the name "Neptune" for the planet. However, in Chinese, Vietnamese, Japanese, and Korean, the planet's name was translated as "sea king star" (海王星), because Neptune was the god of the sea. In Mongolian, Neptune is called *Dalain Van* (Далайн ван), reflecting its namesake god's role as the ruler of the sea. In modern Greek the planet is called *Poseidon* (Ποσειδώνας, *Poseidonas*), the Greek counterpart of Neptune.[8] In Hebrew, "Rahab", (רהב) from a Biblical sea monster mentioned in the Book of Psalms, was selected in a vote managed by the Academy of the Hebrew Language in 2009 as the official name for the planet, even though the existing Latin term "Neptun" (נפטון) is commonly used. In Māori, the planet is called *Tangaroa*, named after the Māori god of the sea.[9] In Nahuatl, the planet is called *Tlāloccītlalli*, named after the rain god Tlāloc. In Thai, Neptune is referred both by its Westernised name *Dao Nepjun* (ดาวเนปจูน), and is also named *Dao Ketu* (ดาวเกตุ, "Star of Ketu"), after the descending lunar node Ketu (केतु) who plays a role in Hindu astrology.

Figure 3: *A size comparison of Neptune and Earth*

Status

From its discovery in 1846 until the subsequent discovery of Pluto in 1930, Neptune was the farthest known planet. When Pluto was discovered, it was considered a planet, and Neptune thus became the second-farthest known planet, except for a 20-year period between 1979 and 1999 when Pluto's elliptical orbit brought it closer than Neptune to the Sun. The discovery of the Kuiper belt in 1992 led many astronomers to debate whether Pluto should be considered a planet or as part of the Kuiper belt. In 2006, the International Astronomical Union defined the word "planet" for the first time, reclassifying Pluto as a "dwarf planet" and making Neptune once again the outermost known planet in the Solar System.

Physical characteristics

Neptune's mass of 1.0243×10^{26} kg is intermediate between Earth and the larger gas giants: it is 17 times that of Earth but just 1/19th that of Jupiter.[10] Its gravity at 1 bar is 11.15 m/s^2, 1.14 times the surface gravity of Earth, and surpassed only by Jupiter. Neptune's equatorial radius of 24,764 km is nearly four times that of Earth. Neptune, like Uranus, is an ice giant, a subclass of giant planet, because they are smaller and have higher concentrations of volatiles

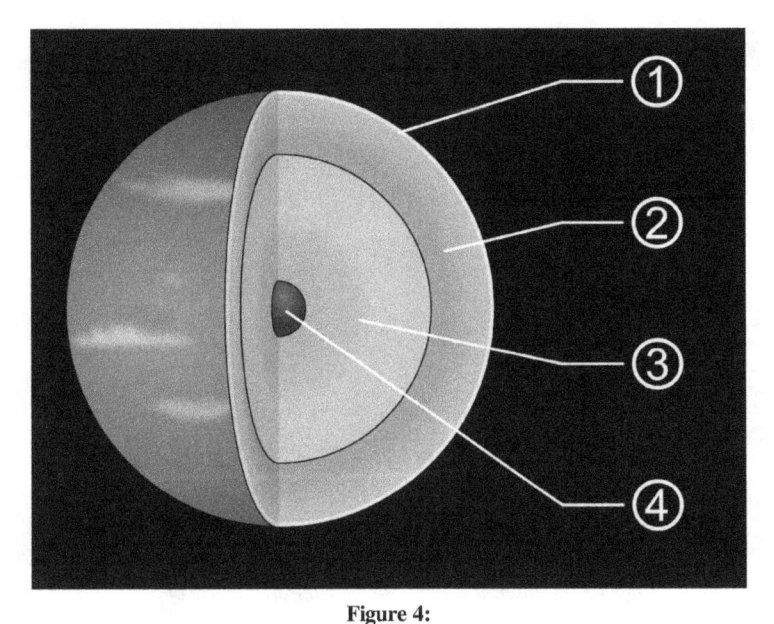

Figure 4:
The internal structure of Neptune:
1. Upper atmosphere, top clouds
2. Atmosphere consisting of hydrogen, helium and methane gas
3. Mantle consisting of water, ammonia and methane ices
4. Core consisting of rock (silicates and nickel–iron)

than Jupiter and Saturn. In the search for extrasolar planets, Neptune has been used as a metonym: discovered bodies of similar mass are often referred to as "Neptunes", just as scientists refer to various extrasolar bodies as "Jupiters".

Internal structure

Neptune's internal structure resembles that of Uranus. Its atmosphere forms about 5% to 10% of its mass and extends perhaps 10% to 20% of the way towards the core, where it reaches pressures of about 10 GPa, or about 100,000 times that of Earth's atmosphere. Increasing concentrations of methane, ammonia and water are found in the lower regions of the atmosphere.

The mantle is equivalent to 10 to 15 Earth masses and is rich in water, ammonia and methane. As is customary in planetary science, this mixture is referred to as icy even though it is a hot, dense fluid. This fluid, which has a high electrical conductivity, is sometimes called a water–ammonia ocean. The mantle may consist of a layer of ionic water in which the water molecules

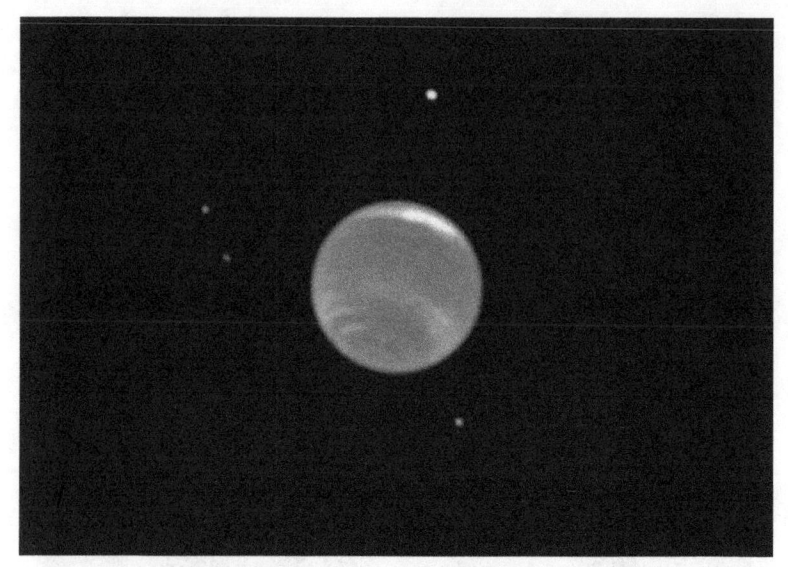

Figure 5: *Combined colour and near-infrared image of Neptune, showing bands of methane in its atmosphere, and four of its moons, Proteus, Larissa, Galatea, and Despina*

break down into a soup of hydrogen and oxygen ions, and deeper down superionic water in which the oxygen crystallises but the hydrogen ions float around freely within the oxygen lattice. At a depth of 7,000 km, the conditions may be such that methane decomposes into diamond crystals that rain downwards like hailstones. Very-high-pressure experiments at the Lawrence Livermore National Laboratory suggest that the base of the mantle may be an ocean of liquid carbon with floating solid 'diamonds'.

The core of Neptune is likely composed of iron, nickel and silicates, with an interior model giving a mass about 1.2 times that of Earth. The pressure at the centre is 7 Mbar (700 GPa), about twice as high as that at the centre of Earth, and the temperature may be 5,400 K.

Atmosphere

At high altitudes, Neptune's atmosphere is 80% hydrogen and 19% helium. A trace amount of methane is also present. Prominent absorption bands of methane exist at wavelengths above 600 nm, in the red and infrared portion of the spectrum. As with Uranus, this absorption of red light by the atmospheric methane is part of what gives Neptune its blue hue, although Neptune's vivid azure differs from Uranus's milder cyan. Because Neptune's atmospheric

Figure 6: *A time-lapse video of Neptune and its moons*

methane content is similar to that of Uranus, some unknown atmospheric constituent is thought to contribute to Neptune's colour.

Neptune's atmosphere is subdivided into two main regions: the lower troposphere, where temperature decreases with altitude, and the stratosphere, where temperature increases with altitude. The boundary between the two, the tropopause, lies at a pressure of 0.1 bars (10 kPa). The stratosphere then gives way to the thermosphere at a pressure lower than 10^{-5} to 10^{-4} bars (1 to 10 Pa). The thermosphere gradually transitions to the exosphere.

Models suggest that Neptune's troposphere is banded by clouds of varying compositions depending on altitude. The upper-level clouds lie at pressures below one bar, where the temperature is suitable for methane to condense. For pressures between one and five bars (100 and 500 kPa), clouds of ammonia and hydrogen sulfide are thought to form. Above a pressure of five bars, the clouds may consist of ammonia, ammonium sulfide, hydrogen sulfide and water. Deeper clouds of water ice should be found at pressures of about 50 bars (5.0 MPa), where the temperature reaches 273 K (0 °C). Underneath, clouds of ammonia and hydrogen sulfide may be found.

High-altitude clouds on Neptune have been observed casting shadows on the opaque cloud deck below. There are also high-altitude cloud bands that wrap around the planet at constant latitude. These circumferential bands have widths

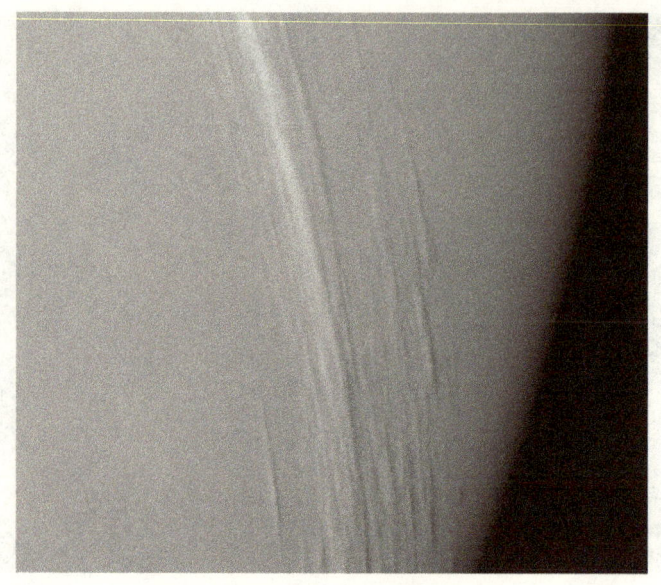

Figure 7: *Bands of high-altitude clouds cast shadows on Neptune's lower cloud deck*

of 50–150 km and lie about 50–110 km above the cloud deck. These altitudes are in the layer where weather occurs, the troposphere. Weather does not occur in the higher stratosphere or thermosphere. Unlike Uranus, Neptune's composition has a higher volume of ocean, whereas Uranus has a smaller mantle.

Neptune's spectra suggest that its lower stratosphere is hazy due to condensation of products of ultraviolet photolysis of methane, such as ethane and ethyne. The stratosphere is also home to trace amounts of carbon monoxide and hydrogen cyanide. The stratosphere of Neptune is warmer than that of Uranus due to the elevated concentration of hydrocarbons.

For reasons that remain obscure, the planet's thermosphere is at an anomalously high temperature of about 750 K. The planet is too far from the Sun for this heat to be generated by ultraviolet radiation. One candidate for a heating mechanism is atmospheric interaction with ions in the planet's magnetic field. Other candidates are gravity waves from the interior that dissipate in the atmosphere. The thermosphere contains traces of carbon dioxide and water, which may have been deposited from external sources such as meteorites and dust.

Magnetosphere

Neptune resembles Uranus in its magnetosphere, with a magnetic field strongly tilted relative to its rotational axis at 47° and offset at least 0.55 radii, or about 13,500 km from the planet's physical centre. Before *Voyager 2*'s arrival at Neptune, it was hypothesised that Uranus's tilted magnetosphere was the result of its sideways rotation. In comparing the magnetic fields of the two planets, scientists now think the extreme orientation may be characteristic of flows in the planets' interiors. This field may be generated by convective fluid motions in a thin spherical shell of electrically conducting liquids (probably a combination of ammonia, methane and water) resulting in a dynamo action.

The dipole component of the magnetic field at the magnetic equator of Neptune is about 14 microteslas (0.14 G). The dipole magnetic moment of Neptune is about 2.2×10^{17} T·m^3 (14 µT·R_N^3, where R_N is the radius of Neptune). Neptune's magnetic field has a complex geometry that includes relatively large contributions from non-dipolar components, including a strong quadrupole moment that may exceed the dipole moment in strength. By contrast, Earth, Jupiter and Saturn have only relatively small quadrupole moments, and their fields are less tilted from the polar axis. The large quadrupole moment of Neptune may be the result of offset from the planet's centre and geometrical constraints of the field's dynamo generator.

Neptune's bow shock, where the magnetosphere begins to slow the solar wind, occurs at a distance of 34.9 times the radius of the planet. The magnetopause, where the pressure of the magnetosphere counterbalances the solar wind, lies at a distance of 23–26.5 times the radius of Neptune. The tail of the magnetosphere extends out to at least 72 times the radius of Neptune, and likely much farther.

Climate

Neptune's weather is characterised by extremely dynamic storm systems, with winds reaching speeds of almost 600 m/s (2,200 km/h; 1,300 mph)—nearly reaching supersonic flow. More typically, by tracking the motion of persistent clouds, wind speeds have been shown to vary from 20 m/s in the easterly direction to 325 m/s westward. At the cloud tops, the prevailing winds range in speed from 400 m/s along the equator to 250 m/s at the poles. Most of the winds on Neptune move in a direction opposite the planet's rotation.[11] The general pattern of winds showed prograde rotation at high latitudes vs. retrograde rotation at lower latitudes. The difference in flow direction is thought to be a "skin effect" and not due to any deeper atmospheric processes. At 70° S latitude, a high-speed jet travels at a speed of 300 m/s.

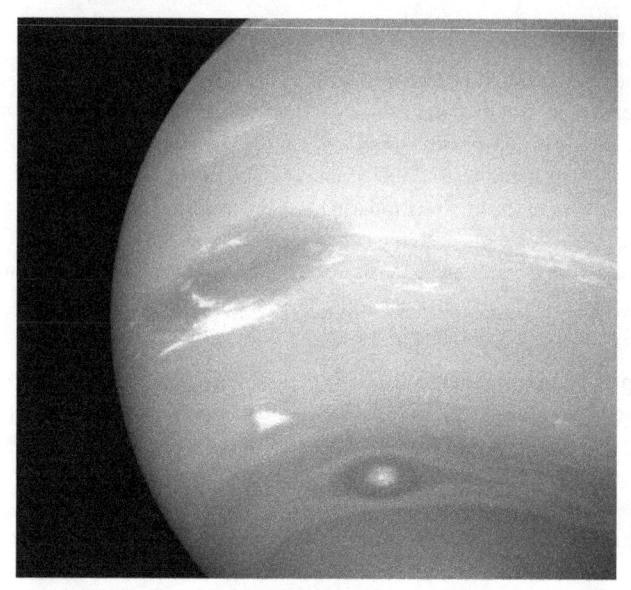

Figure 8: *The Great Dark Spot (top), Scooter (middle white cloud), and the Small Dark Spot (bottom), with contrast exaggerated.*

Neptune differs from Uranus in its typical level of meteorological activity. *Voyager 2* observed weather phenomena on Neptune during its 1989 flyby, but no comparable phenomena on Uranus during its 1986 fly-by.

The abundance of methane, ethane and acetylene at Neptune's equator is 10–100 times greater than at the poles. This is interpreted as evidence for upwelling at the equator and subsidence near the poles.Wikipedia:Please clarify

In 2007, it was discovered that the upper troposphere of Neptune's south pole was about 10 K warmer than the rest of its atmosphere, which averages approximately 73 K (–200 °C). The temperature differential is enough to let methane, which elsewhere is frozen in the troposphere, escape into the stratosphere near the pole. The relative "hot spot" is due to Neptune's axial tilt, which has exposed the south pole to the Sun for the last quarter of Neptune's year, or roughly 40 Earth years. As Neptune slowly moves towards the opposite side of the Sun, the south pole will be darkened and the north pole illuminated, causing the methane release to shift to the north pole.

Because of seasonal changes, the cloud bands in the southern hemisphere of Neptune have been observed to increase in size and albedo. This trend was

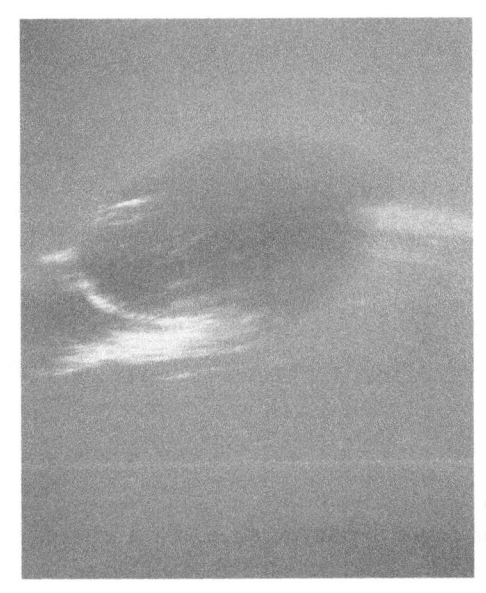

Figure 9: *The Great Dark Spot, as imaged by Voyager 2*

first seen in 1980 and is expected to last until about 2020. The long orbital period of Neptune results in seasons lasting forty years.

Storms

In 1989, the Great Dark Spot, an anti-cyclonic storm system spanning 13,000 × 6,600 km, was discovered by NASA's *Voyager 2* spacecraft. The storm resembled the Great Red Spot of Jupiter. Some five years later, on 2 November 1994, the *Hubble Space Telescope* did not see the Great Dark Spot on the planet. Instead, a new storm similar to the Great Dark Spot was found in Neptune's northern hemisphere.

The Scooter is another storm, a white cloud group farther south than the Great Dark Spot. This nickname first arose during the months leading up to the *Voyager 2* encounter in 1989, when they were observed moving at speeds faster than the Great Dark Spot (and images acquired later would subsequently reveal the presence of clouds moving even faster than those that had initially been detected by *Voyager 2*). The Small Dark Spot is a southern cyclonic storm, the second-most-intense storm observed during the 1989 encounter. It was initially completely dark, but as *Voyager 2* approached the planet, a bright core developed and can be seen in most of the highest-resolution images.

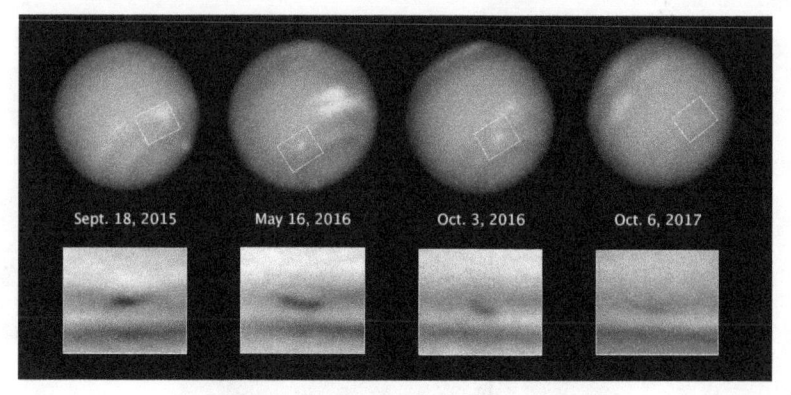

Figure 10: *Neptune's shrinking vortex.*

Neptune's dark spots are thought to occur in the troposphere at lower altitudes than the brighter cloud features, so they appear as holes in the upper cloud decks. As they are stable features that can persist for several months, they are thought to be vortex structures. Often associated with dark spots are brighter, persistent methane clouds that form around the tropopause layer. The persistence of companion clouds shows that some former dark spots may continue to exist as cyclones even though they are no longer visible as a dark feature. Dark spots may dissipate when they migrate too close to the equator or possibly through some other unknown mechanism.

Internal heating

Neptune's more varied weather when compared to Uranus is due in part to its higher internal heating. Although Neptune lies over 50% farther from the Sun than Uranus, and receives only 40% its amount of sunlight, the two planets' surface temperatures are roughly equal. The upper regions of Neptune's troposphere reach a low temperature of 51.8 K (–221.3 °C). At a depth where the atmospheric pressure equals 1 bar (100 kPa), the temperature is 72.00 K (–201.15 °C). Deeper inside the layers of gas, the temperature rises steadily. As with Uranus, the source of this heating is unknown, but the discrepancy is larger: Uranus only radiates 1.1 times as much energy as it receives from the Sun; whereas Neptune radiates about 2.61 times as much energy as it receives from the Sun. Neptune is the farthest planet from the Sun, yet its internal energy is sufficient to drive the fastest planetary winds seen in the Solar System. Depending on the thermal properties of its interior, the heat left over from Neptune's formation may be sufficient to explain its current heat flow, though it is more difficult to simultaneously explain Uranus's lack of internal heat while preserving the apparent similarity between the two planets.[12]

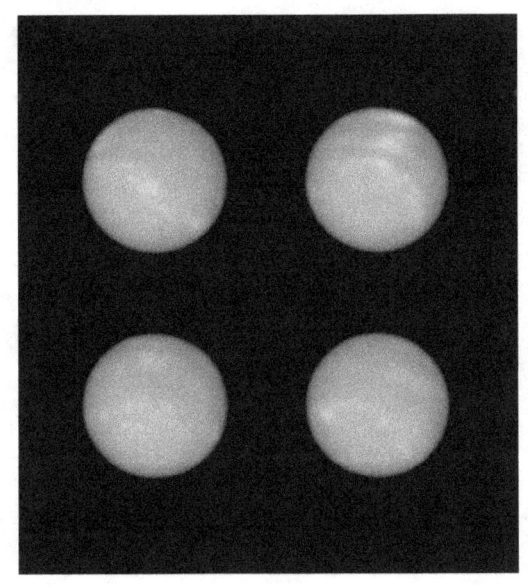

Figure 11: *Four images taken a few hours apart with the NASA/ESA Hubble Space Telescope's Wide Field Camera 3*

Orbit and rotation

The average distance between Neptune and the Sun is 4.5 billion km (about 30.1 astronomical units (AU)), and it completes an orbit on average every 164.79 years, subject to a variability of around ±0.1 years. The perihelion distance is 29.81 AU; the aphelion distance is 30.33 AU.[13]

On 11 July 2011, Neptune completed its first full barycentric orbit since its discovery in 1846, although it did not appear at its exact discovery position in the sky, because Earth was in a different location in its 365.26-day orbit. Because of the motion of the Sun in relation to the barycentre of the Solar System, on 11 July Neptune was also not at its exact discovery position in relation to the Sun; if the more common heliocentric coordinate system is used, the discovery longitude was reached on 12 July 2011.[14,15]

The elliptical orbit of Neptune is inclined 1.77° compared to that of Earth.

The axial tilt of Neptune is 28.32°, which is similar to the tilts of Earth (23°) and Mars (25°). As a result, Neptune experiences similar seasonal changes to Earth. The long orbital period of Neptune means that the seasons last for forty Earth years. Its sidereal rotation period (day) is roughly 16.11 hours. Because

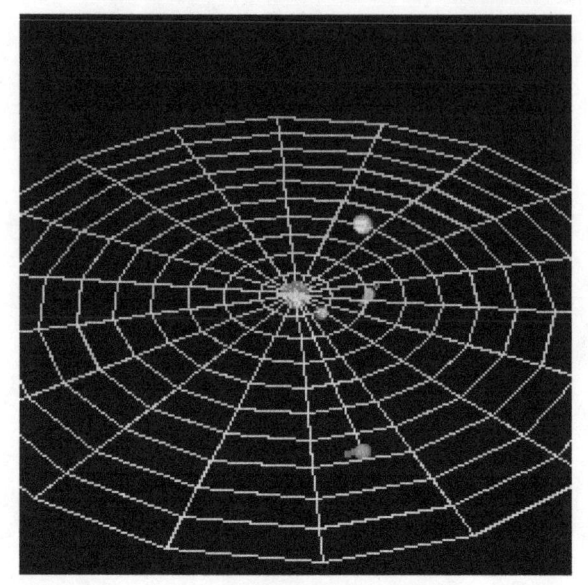

Figure 12: *Neptune (red arc) completes one orbit around the Sun (centre) for every 164.79 orbits of Earth. The light blue object represents Uranus.*

its axial tilt is comparable to Earth's, the variation in the length of its day over the course of its long year is not any more extreme.

Because Neptune is not a solid body, its atmosphere undergoes differential rotation. The wide equatorial zone rotates with a period of about 18 hours, which is slower than the 16.1-hour rotation of the planet's magnetic field. By contrast, the reverse is true for the polar regions where the rotation period is 12 hours. This differential rotation is the most pronounced of any planet in the Solar System, and it results in strong latitudinal wind shear.

Orbital resonances

Neptune's orbit has a profound impact on the region directly beyond it, known as the Kuiper belt. The Kuiper belt is a ring of small icy worlds, similar to the asteroid belt but far larger, extending from Neptune's orbit at 30 AU out to about 55 AU from the Sun. Much in the same way that Jupiter's gravity dominates the asteroid belt, shaping its structure, so Neptune's gravity dominates the Kuiper belt. Over the age of the Solar System, certain regions of the Kuiper belt became destabilised by Neptune's gravity, creating gaps in the Kuiper belt's structure. The region between 40 and 42 AU is an example.

Kuiper belt and orbital resonance

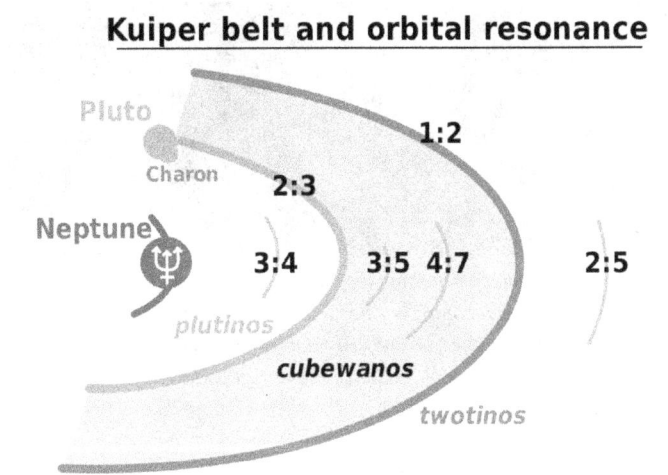

Figure 13: *A diagram showing the major orbital resonances in the Kuiper belt caused by Neptune: the highlighted regions are the 2:3 resonance (plutinos), the nonresonant "classical belt" (cubewanos), and the 1:2 resonance (twotinos).*

There do exist orbits within these empty regions where objects can survive for the age of the Solar System. These resonances occur when Neptune's orbital period is a precise fraction of that of the object, such as 1:2, or 3:4. If, say, an object orbits the Sun once for every two Neptune orbits, it will only complete half an orbit by the time Neptune returns to its original position. The most heavily populated resonance in the Kuiper belt, with over 200 known objects, is the 2:3 resonance. Objects in this resonance complete 2 orbits for every 3 of Neptune, and are known as plutinos because the largest of the known Kuiper belt objects, Pluto, is among them. Although Pluto crosses Neptune's orbit regularly, the 2:3 resonance ensures they can never collide. The 3:4, 3:5, 4:7 and 2:5 resonances are less populated.

Neptune has a number of known trojan objects occupying both the Sun–Neptune L_4 and L_5 Lagrangian points—gravitationally stable regions leading and trailing Neptune in its orbit, respectively. Neptune trojans can be viewed as being in a 1:1 resonance with Neptune. Some Neptune trojans are remarkably stable in their orbits, and are likely to have formed alongside Neptune rather than being captured. The first object identified as associated with Neptune's trailing L_5 Lagrangian point was 2008 LC$_{18}$. Neptune also has a temporary quasi-satellite, (309239) 2007 RW$_{10}$. The object has been

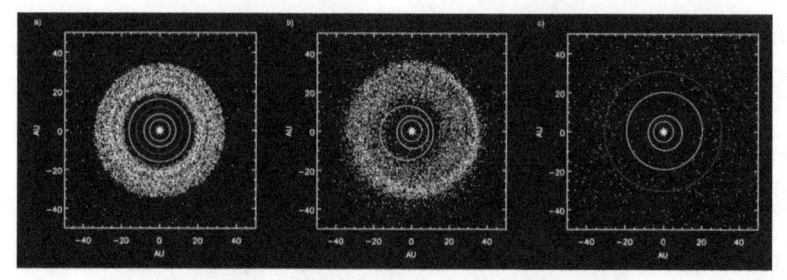

Figure 14: *A simulation showing the outer planets and Kuiper belt: a) before Jupiter and Saturn reached a 2:1 resonance; b) after inward scattering of Kuiper belt objects following the orbital shift of Neptune; c) after ejection of scattered Kuiper belt bodies by Jupiter*

a quasi-satellite of Neptune for about 12,500 years and it will remain in that dynamical state for another 12,500 years.

Formation and migration

The formation of the ice giants, Neptune and Uranus, has proven difficult to model precisely. Current models suggest that the matter density in the outer regions of the Solar System was too low to account for the formation of such large bodies from the traditionally accepted method of core accretion, and various hypotheses have been advanced to explain their formation. One is that the ice giants were not formed by core accretion but from instabilities within the original protoplanetary disc and later had their atmospheres blasted away by radiation from a nearby massive OB star.

An alternative concept is that they formed closer to the Sun, where the matter density was higher, and then subsequently migrated to their current orbits after the removal of the gaseous protoplanetary disc. This hypothesis of migration after formation is favoured, due to its ability to better explain the occupancy of the populations of small objects observed in the trans-Neptunian region. The current most widely accepted explanation of the details of this hypothesis is known as the Nice model, which explores the effect of a migrating Neptune and the other giant planets on the structure of the Kuiper belt.

Moons

Neptune has 14 known moons.[16] Triton is the largest Neptunian moon, comprising more than 99.5% of the mass in orbit around Neptune,[17] and it is the only one massive enough to be spheroidal. Triton was discovered by William

Figure 15: *Natural-colour view of Neptune with Proteus (top), Larissa (lower right), and Despina (left), from the Hubble Space Telescope*

Lassell just 17 days after the discovery of Neptune itself. Unlike all other large planetary moons in the Solar System, Triton has a retrograde orbit, indicating that it was captured rather than forming in place; it was probably once a dwarf planet in the Kuiper belt. It is close enough to Neptune to be locked into a synchronous rotation, and it is slowly spiralling inward because of tidal acceleration. It will eventually be torn apart, in about 3.6 billion years, when it reaches the Roche limit. In 1989, Triton was the coldest object that had yet been measured in the Solar System, with estimated temperatures of 38 K (−235 °C).

Neptune's second known satellite (by order of discovery), the irregular moon Nereid, has one of the most eccentric orbits of any satellite in the Solar System. The eccentricity of 0.7512 gives it an apoapsis that is seven times its periapsis distance from Neptune.[18]

From July to September 1989, *Voyager 2* discovered six moons of Neptune. Of these, the irregularly shaped Proteus is notable for being as large as a body of its density can be without being pulled into a spherical shape by its own gravity. Although the second-most-massive Neptunian moon, it is only 0.25% the mass of Triton. Neptune's innermost four moons—Naiad, Thalassa, Despina and Galatea—orbit close enough to be within Neptune's rings. The next-farthest

Figure 16: *Neptune's moon Proteus*

out, Larissa, was originally discovered in 1981 when it had occulted a star. This occultation had been attributed to ring arcs, but when *Voyager 2* observed Neptune in 1989, Larissa was found to have caused it. Five new irregular moons discovered between 2002 and 2003 were announced in 2004. A new moon and the smallest yet, S/2004 N 1, was found in 2013. Because Neptune was the Roman god of the sea, Neptune's moons have been named after lesser sea gods.

Planetary rings

Neptune has a planetary ring system, though one much less substantial than that of Saturn. The rings may consist of ice particles coated with silicates or carbon-based material, which most likely gives them a reddish hue. The three main rings are the narrow Adams Ring, 63,000 km from the centre of Neptune, the Le Verrier Ring, at 53,000 km, and the broader, fainter Galle Ring, at 42,000 km. A faint outward extension to the Le Verrier Ring has been named Lassell; it is bounded at its outer edge by the Arago Ring at 57,000 km.

The first of these planetary rings was detected in 1968 by a team led by Edward Guinan. In the early 1980s, analysis of this data along with newer observations led to the hypothesis that this ring might be incomplete. Evidence that the rings might have gaps first arose during a stellar occultation in 1984 when the rings

Figure 17: *Neptune's rings*

obscured a star on immersion but not on emersion. Images from *Voyager 2* in 1989 settled the issue by showing several faint rings.

The outermost ring, Adams, contains five prominent arcs now named *Courage*, *Liberté*, *Egalité 1*, *Egalité 2* and *Fraternité* (Courage, Liberty, Equality and Fraternity). The existence of arcs was difficult to explain because the laws of motion would predict that arcs would spread out into a uniform ring over short timescales. Astronomers now estimate that the arcs are corralled into their current form by the gravitational effects of Galatea, a moon just inward from the ring.

Earth-based observations announced in 2005 appeared to show that Neptune's rings are much more unstable than previously thought. Images taken from the W. M. Keck Observatory in 2002 and 2003 show considerable decay in the rings when compared to images by *Voyager 2*. In particular, it seems that the *Liberté* arc might disappear in as little as one century.

Observation

With an apparent magnitude between +7.7 and +8.0, Neptune is never visible to the naked eye and can be outshone by Jupiter's Galilean moons, the dwarf planet Ceres and the asteroids 4 Vesta, 2 Pallas, 7 Iris, 3 Juno, and 6 Hebe.[19] A

Figure 18: *In 2018, the European Southern Observatory developed unique laser-based methods to get clear and high-resolution images of Neptune from the surface of Earth.*

telescope or strong binoculars will resolve Neptune as a small blue disk, similar in appearance to Uranus.[20]

Because of the distance of Neptune from Earth, its angular diameter only ranges from 2.2 to 2.4 arcseconds, the smallest of the Solar System planets. Its small apparent size makes it challenging to study it visually. Most telescopic data was fairly limited until the advent of the *Hubble Space Telescope* and large ground-based telescopes with adaptive optics (AO).[21] The first scientifically useful observation of Neptune from ground-based telescopes using adaptive optics, was commenced in 1997 from Hawaii.[22] Neptune is currently entering its spring and summer season and has been shown to be heating up, with increased atmospheric activity and brightness as a consequence. Combined with technological advancements, ground-based telescopes with adaptive optics are recording increasingly more detailed images of it. Both *Hubble* and the adaptive-optics telescopes on Earth have made many new discoveries within the Solar System since the mid-1990s, with a large increase in the number of known satellites and moons around the outer planet, among others. In 2004 and 2005, five new small satellites of Neptune with diameters between 38 and 61 kilometres were discovered.[23]

Figure 19: *A Voyager 2 mosaic of Triton*

From Earth, Neptune goes through apparent retrograde motion every 367 days, resulting in a looping motion against the background stars during each opposition. These loops carried it close to the 1846 discovery coordinates in April and July 2010 and again in October and November 2011.

Observation of Neptune in the radio-frequency band shows that it is a source of both continuous emission and irregular bursts. Both sources are thought to originate from its rotating magnetic field. In the infrared part of the spectrum, Neptune's storms appear bright against the cooler background, allowing the size and shape of these features to be readily tracked.

Exploration

Voyager 2 is the only spacecraft that has visited Neptune. The spacecraft's closest approach to the planet occurred on 25 August 1989. Because this was the last major planet the spacecraft could visit, it was decided to make a close flyby of the moon Triton, regardless of the consequences to the trajectory, similarly to what was done for *Voyager 1*'s encounter with Saturn and its moon Titan. The images relayed back to Earth from *Voyager 2* became the basis of a 1989 PBS all-night program, *Neptune All Night*.

During the encounter, signals from the spacecraft required 246 minutes to reach Earth. Hence, for the most part, *Voyager 2*'s mission relied on preloaded commands for the Neptune encounter. The spacecraft performed a near-encounter with the moon Nereid before it came within 4,400 km of Neptune's atmosphere on 25 August, then passed close to the planet's largest moon Triton later the same day.[24]

The spacecraft verified the existence of a magnetic field surrounding the planet and discovered that the field was offset from the centre and tilted in a manner similar to the field around Uranus. Neptune's rotation period was determined using measurements of radio emissions and *Voyager 2* also showed that Neptune had a surprisingly active weather system. Six new moons were discovered, and the planet was shown to have more than one ring.

The flyby also provided the first accurate measurement of Neptune's mass which was found to be 0.5 percent less than previously calculated. The new figure disproved the hypothesis that an undiscovered Planet X acted upon the orbits of Neptune and Uranus.[25]

After the *Voyager 2* flyby mission, the next step in scientific exploration of the Neptunian system, is considered to be a Flagship orbital mission. Such a hypothetical mission is envisioned to be possible in the late 2020s or early 2030s. However, there have been discussions to launch Neptune missions sooner. In 2003, there was a proposal in NASA's "Vision Missions Studies" for a "Neptune Orbiter with Probes" mission that does *Cassini*-level science. Another, more recent proposal was for *Argo*, a flyby spacecraft to be launched in 2019, that would visit Jupiter, Saturn, Neptune, and a Kuiper belt object. The focus would be on Neptune and its largest moon Triton to be investigated around 2029. The proposed *New Horizons 2* mission (which was later scrapped) might also have done a close flyby of the Neptunian system.

Bibliography

- Burgess, Eric (1991). *Far Encounter: The Neptune System*. Columbia University Press. ISBN 978-0-231-07412-4.
- Moore, Patrick (2000). *The Data Book of Astronomy*. CRC Press. ISBN 978-0-7503-0620-1.

Further reading

- Miner, Ellis D.; Wessen, Randii R. (2002). *Neptune: The Planet, Rings, and Satellites*. Springer-Verlag. ISBN 978-1-85233-216-7.
- Standage, Tom (2001). *The Neptune File*. Penguin. ISBN 978-0-8027-1363-6.

External links

- NASA's Neptune fact sheet[26]
- Neptune[27] from Bill Arnett's nineplanets.org
- Neptune[28] Astronomy Cast episode No. 63, includes full transcript.
- Neptune Profile[29] at NASA's Solar System Exploration site[30]
- Planets – Neptune[31] A children's guide to Neptune.
- Merrifield, Michael; Bauer, Amanda (2010). "Neptune"[32]. *Sixty Symbols*. Brady Haran for the University of Nottingham.
- Neptune by amateur[33] (The Planetary Society)

<indicator name="featured-star"> ★ </indicator>

Discovery

Discovery of Neptune

The planet Neptune was mathematically predicted before it was directly observed. With a prediction by Urbain Le Verrier, telescopic observations confirming the existence of a major planet were made on the night of September 23–24, 1846, at the Berlin Observatory, by astronomer Johann Gottfried Galle (assisted by Heinrich Louis d'Arrest), working from Le Verrier's calculations. It was a sensational moment of 19th-century science, and dramatic confirmation of Newtonian gravitational theory. In François Arago's apt phrase, Le Verrier had discovered a planet "with the point of his pen".

In retrospect, after it was discovered, it turned out it had been observed many times before but not recognized, and there were others who made various calculations about its location which did not lead to its observation. By 1847, the planet Uranus had completed nearly one full orbit since its discovery by William Herschel in 1781, and astronomers had detected a series of irregularities in its path that could not be entirely explained by Newton's law of gravitation. These irregularities could, however, be resolved if the gravity of a farther, unknown planet were disturbing its path around the Sun. In 1845, astronomers Urbain Le Verrier in Paris and John Couch Adams in Cambridge separately began calculations to determine the nature and position of such a planet. Le Verrier's success also led to a tense international dispute over priority, because shortly after the discovery George Airy, at the time British Astronomer Royal, announced that Adams had also predicted the discovery of the planet. Nevertheless, the Royal Society awarded Le Verrier the Copley medal in 1846 for his achievement, without mention of Adams.

The discovery of Neptune led to the discovery of its moon Triton by William Lassell just seventeen days later.

Figure 20: *New Berlin Observatory at Linden Street, where Neptune was discovered observationally.*

Earlier observations

Neptune is too dim to be visible to the naked eye: its apparent magnitude is never brighter than 7.7. Therefore, the first observations of Neptune were only possible after the invention of the telescope. There is evidence that Neptune was seen and recorded by Galileo Galilei in 1613, Jérôme Lalande in 1795 and John Herschel in 1830, but none is known to have recognized it as a planet at the time. These pre-discovery observations were important in accurately determining the orbit of Neptune. Neptune would appear prominently even in early telescopes so other pre-discovery observation records are likely.

Galileo's drawings show that he observed Neptune on December 28, 1612, and again on January 27, 1613; on both occasions, Galileo mistook Neptune for a fixed star when it appeared very close (in conjunction) to Jupiter in the night sky. Historically it was thought that he believed it to be a fixed blue star, and so he is not credited with its discovery. At the time of his first observation in December 1612, it was stationary in the sky because it had just turned retrograde that very day; because it was only beginning its yearly retrograde cycle, Neptune's motion was thought to be too slight, and its apparent size too small, to clearly appear to be a planet in Galileo's small telescope. However, in July 2009 University of Melbourne physicist David Jamieson announced new evidence suggesting that Galileo was indeed aware that he had discovered something unusual about this "star". Galileo, in one of his notebooks, noted the movement of a background star (Neptune) on January 28 and a dot (in Neptune's position) drawn in a different ink suggests that he found it on an earlier sketch, drawn on the night of January 6, suggesting a systematic search among his earlier observations. However, so far there is neither clear evidence that he identified this moving object as a planet, nor that he published these observations of it. There is no evidence that he ever attempted to observe it again.

In 1847, Sears C. Walker of the U.S. Naval Observatory searched historical records and surveys for possible prediscovery sightings of the planet Neptune. He found that observations made by Lalande's staff at the Paris Observatory in 1795 were in the direction of Neptune's position in the sky. In the catalog observations for May 8 and again on May 10 of 1795 a *star* was observed in the approximate position expected for Neptune. The uncertainty of the position was noted with a colon. This notation was also used to indicate an observation error so it was not until the original records of the observatory were reviewed that it was established with certainty that the object was Neptune and the position error in the observations made two nights apart was due to the planet's motion across the sky. The discovery of these records of Neptune's position in 1795 led to a better calculation of the planet's orbit.

John Herschel almost discovered Neptune the same way his father, William Herschel, had discovered Uranus in 1781, by chance observation. In an 1846 letter to Wilhelm Struve, John Herschel states that he observed Neptune during a sweep of the sky on July 14, 1830. Although his telescope was powerful enough to resolve Neptune into a small blue disk and show it to be a planet, he did not recognize it at the time and mistook it for a star.

Irregularities in Uranus's orbit

In 1821, Alexis Bouvard had published astronomical tables of the orbit of Uranus, making predictions of future positions based on Newton's laws of motion and gravitation.[34] Subsequent observations revealed substantial deviations from the tables, leading Bouvard to hypothesize some perturbing body.[35] These irregularities or "residuals", both in the planet's ecliptic longitude and in its distance from the Sun, or radius vector, might be explained by a number of hypotheses: the effect of the Sun's gravity, at such a great distance might differ from Newton's description; or the discrepancies might simply be observational error; or perhaps Uranus was being pulled, or perturbed, by an as-yet undiscovered planet.

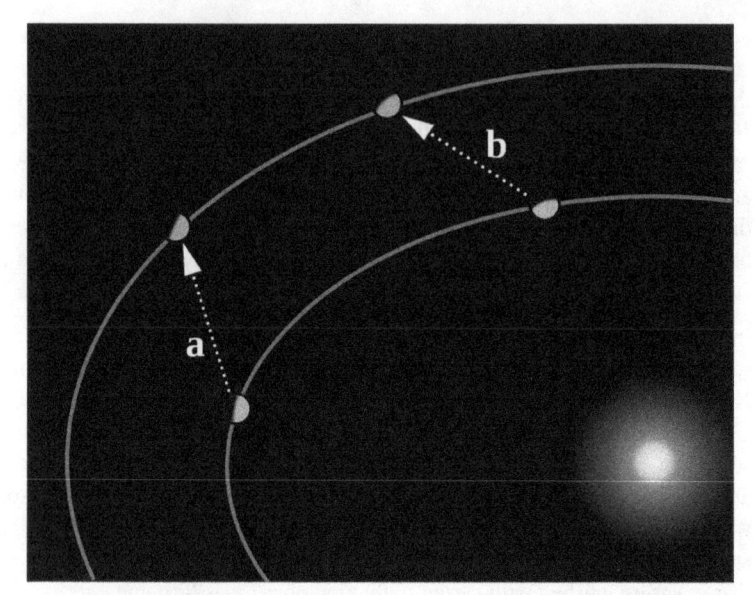

Figure 21: *At position a, Neptune gravitationally perturbs the orbit of Uranus, pulling it ahead of the predicted location. The reverse is true at b, where the perturbation retards the orbital motion of Uranus.*

Adams learned of the irregularities while still an undergraduate and became convinced of the "perturbation" hypothesis. Adams believed, in the face of anything that had been attempted before, that he could use the observed data on Uranus, and utilising nothing more than Newton's law of gravitation, deduce the mass, position and orbit of the perturbing body.

After his final examinations in 1843, Adams was elected fellow of his college and spent the summer vacation in Cornwall calculating the first of six iterations.

In modern terms, the problem is an inverse problem, an attempt to deduce the parameters of a mathematical model from observed data. Though the problem is a simple one for modern mathematics after the advent of electronic computers, at the time it involved much laborious hand calculation. Adams began by assuming a nominal position for the hypothesised body, using the empirical Bode's law. He then calculated the path of Uranus using the assumed position of the perturbing body and calculated the difference between his calculated path and the observations, in modern terms the residuals. He then adjusted the characteristics of the perturbing body in a way suggested by the residuals and repeated the process, a process similar to regression analysis.

Figure 22: *John Couch Adams*

On 13 February 1844, James Challis, director of the Cambridge Observatory, requested data on the position of Uranus, for Adams, from Astronomer Royal George Biddell Airy at the Royal Observatory, Greenwich. Adams certainly completed some calculations on 18 September 1845.

Supposedly, Adams communicated his work to Challis in mid-September 1845 but there is some controversy as to how. The story and date of this communication only seem to have come to light in a letter from Challis to the *Athenaeum* dated 17 October 1846. However, no document was identified until 1904 when Sampson suggested a note in Adams's papers that describes "the New Planet" and is endorsed, in handwriting not Adams's, with the note "Received in September 1845".[36] Though this has often been taken to establish Adams's priority, some historians have disputed its authenticity, on the basis that "the New Planet" was not a term current in 1845, and on the basis that the note is dated only after the fact by someone other than Adams. Further, the results of the calculations are different from those communicated to Airy a few weeks later. Adams certainly gave Challis no detailed calculations and Challis was unimpressed by the description of his method of successively approximating the position of the body, being disinclined to start a laborious observational programme at the observatory, remarking "while the labour was certain, success appeared to be so uncertain."

Figure 23: *Urbain Jean-Joseph Le Verrier.*

Meanwhile, Urbain Le Verrier, on November 10, 1845, presented to the *Académie des sciences* in Paris a memoir on Uranus, showing that the preexisting theory failed to account for its motion. Unaware of Adams's work, he attempted a similar investigation, and on June 1, 1846, in a second memoir presented to a public meeting of the Académie, gave the position, but not the mass or orbit, of the proposed perturbing body. Le Verrier located Neptune within one degree of its discovery position.

The search

Upon receiving in England the news of Le Verrier's June prediction, George Airy immediately recognized the similarity of Le Verrier's and Adams' solutions. Up until that moment, Adams' work had been little more than a curiosity, but independent confirmation from Le Verrier spurred Airy to organize a secret attempt to find the planet.[37,38] At a July 1846 meeting of the Board of Visitors of the Greenwich Observatory, with Challis and Sir John Herschel present, Airy suggested that Challis urgently look for the planet with the Cambridge 11.25 inch equatorial telescope, "in the hope of rescuing the matter from a state which is ... almost desperate".[39] The search was begun by a laborious method on 29 July. Adams continued to work on the problem, providing the British team with six solutions in 1845 and 1846[40] which

Figure 24: *Johann Gottfried Galle, 1880*

sent Challis searching the wrong part of the sky. Only after the discovery of Neptune had been announced in Paris and Berlin did it become apparent that Neptune had been observed on August 8 and August 12 but because Challis lacked an up-to-date star-map, it was not recognized as a planet.

Discovery observation: September 24, 1846

Le Verrier was unaware that his public confirmation of Adams' private computations had set in motion a British search for the purported planet. On 31 August, Le Verrier presented a third memoir, now giving the mass and orbit of the new body. Having been unsuccessful in his efforts to interest any French astronomer in the problem, Le Verrier finally sent his results by post to Johann Gottfried Galle at the Berlin Observatory. Galle received Le Verrier's letter on 23 September and immediately set to work observing in the region suggested by Le Verrier. Galle's student, Heinrich Louis d'Arrest, suggested that a recently drawn chart of the sky, in the region of Le Verrier's predicted location, could be compared with the current sky to seek the displacement characteristic of a planet, as opposed to a stationary star.

Neptune was discovered just after midnight, after less than an hour of searching and less than 1 degree from the position Le Verrier had predicted, a remarkable match. After two further nights of observations in which its position and movement were verified, Galle replied to Le Verrier with astonishment: "the planet whose place you have [computed] *really exists*" (emphasis in original). The discovery telescope was an equatorial mounted achromatic refractor by Joseph Fraunhofer's firm Merz und Mahler.[41]

Aftermath

On the announcement of the discovery, Herschel, Challis and Richard Sheepshanks, foreign secretary of the Royal Astronomical Society, announced that Adams had already calculated the planet's characteristics and position. Airy, at length, published an account of the circumstances, and Adams's memoir was printed as an appendix to the *Nautical Almanac*. However, it appears that the version published by Airy had been edited by the omission of a "crucial phrase" to disguise the fact that Adams had quoted only mean longitude and not the orbital elements.

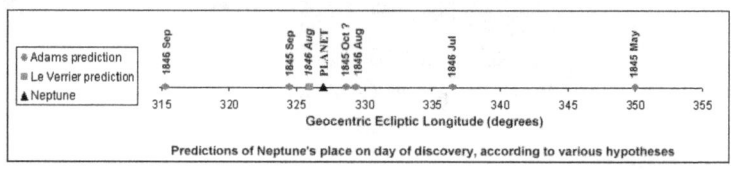

● Adams prediction ■ Le Verrier prediction ▲ Neptune	1846 Sep	1845 Sep 1846 Aug PLANET 1845 Oct ? 1846 Aug	1846 Jul	1845 May
	315 320	325 330 335	340 345	350 355
		Geocentric Ecliptic Longitude (degrees)		
	Predictions of Neptune's place on day of discovery, according to various hypotheses			

A keen controversy arose in France and England as to the merits of the two astronomers. There was much criticism of Airy in England. Adams was a diffident young man who was naturally reluctant to publish a result that would establish or ruin his career. Airy and Challis were criticised, particularly by James Glaisher, as failing to exercise their proper role as mentors of a young talent. Challis was contrite but Airy defended his own behaviour, claiming that the search for a planet was not the role of the Greenwich Observatory. On the whole, Airy has been defended by his biographers. In France the claims made for an unknown Englishman were resented as detracting from the credit due to Le Verrier's achievement.

The Royal Society awarded Le Verrier the Copley medal in 1846 for his achievement, without mention of Adams, but Adams's academic reputation at Cambridge, and in society, was assured. As the facts became known, some British astronomers pushed the view that the two astronomers had independently solved the problem of Neptune, and ascribed equal importance to each. But Adams himself publicly acknowledged Le Verrier's priority and credit (not forgetting to mention the role of Galle) in the paper that he gave to the Royal Astronomical Society in November 1846:

<templatestyles src="Template:Quote/styles.css"/>

> *I mention these dates merely to show that my results were arrived at independently, and previously to the publication of those of M. Le Verrier, and not with the intention of interfering with his just claims to the honours of the discovery ; for there is no doubt that his researches were first published to the world, and led to the actual discovery of the planet by Dr. Galle, so that the facts stated above cannot detract, in the slightest degree, from the credit due to M. Le Verrier.*
>
> *—Adams (1846)*

The criticism was soon afterwards made, that both Adams and Le Verrier had been over-optimistic in the precision they claimed for their calculations, and both had, by using Bode's law, greatly overestimated the planet's distance from the sun. Further, it was suggested that they both succeeded in getting the longitude almost right only because of a "fluke of orbital timing". This criticism was discussed in detail by Danjon (1946) who illustrated with a diagram and discussion that while hypothetical orbits calculated by both Le Verrier and Adams for the new planet were indeed of very different size on the whole from that of the real Neptune (and actually similar to each other), they were both much closer to the real Neptune over that crucial segment of orbit covering the interval of years for which the observations and calculations were made, than they were for the rest of the calculated orbits. So the fact that both the calculators used a much larger orbital major axis than the reality was shown to be not so important, and not the most relevant parameter.

The new planet, at first called "Le Verrier" by François Arago, received by consensus the neutral name of Neptune. Its mathematical prediction was a great intellectual feat, but it showed also that Newton's law of gravitation, which Airy had almost called in question, prevailed even at the limits of the solar system.

Adams held no bitterness towards Challis or Airy and acknowledged his own failure to convince the astronomical world:

<templatestyles src="Template:Quote/styles.css"/>

> *I could not expect however that practical astronomers, who were already fully occupied with important labours, would feel as much confidence in the results of my investigations, as I myself did.*

By contrast, Le Verrier was arrogant and assertive, enabling the British scientific establishment to close ranks behind Adams while the French, in general, found little sympathy with Le Verrier. In 1874–1876, Adams was President of the Royal Astronomical Society when it fell to him to present the RAS Gold Medal of the year to Le Verrier.

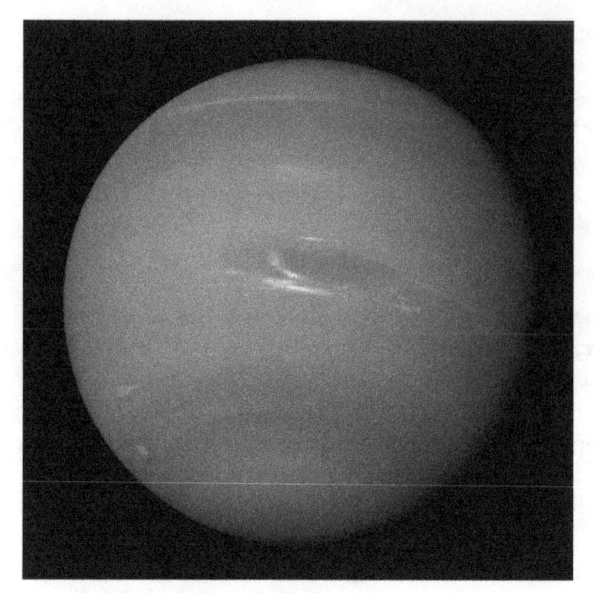

Figure 25: *Neptune in 1989 by the Voyager 2 probe*

Later analysis

The conventional wisdom that Neptune's discovery should be "credited to both Adams and Le Verrier" has recently been challenged putting in doubt the accounts of Airy, Challis and Adams in 1846.

In 1999, Adams's correspondence with Airy, which had been lost by the Royal Greenwich Observatory, was rediscovered in Chile among the possessions of astronomer Olin J. Eggen after his death. In an interview in 2003, historian Nicholas Kollerstrom concluded that Adams's claim to Neptune was far weaker than had been suggested, as he had vacillated repeatedly over the planet's exact location, with estimates ranging across 20 degrees of arc. Airy's role as the hidebound superior willfully ignoring the upstart young intellect was, according to Kollerstrom, largely constructed after the planet was found, in order to boost Adams's, and therefore Britain's, credit for the discovery. A later *Scientific American* article by Sheehan, Kollerstrom and Waff claimed more boldly "The Brits Stole Neptune" and concluded "The achievement was Le Verrier's alone."

Neptune discovery telescope

The telescope, at New Berlin Observatory (1835–1913), that discovered Neptune was an achromatic refractor with an aperture of 9 Paris inches (9.6 English inches, or 24.4 cm). Made by the late Joseph Fraunhofer's firm, Merz und Mahler, it was a high-performance telescope of its era, with one of the largest achromatic doublets available and a finely made equatorial mount, with a clock drive to move the 4 m (13.4') main tube at the same rate as Earth's rotation. Eventually the telescope was moved to the Deutsches Museum in Munich, Germany, where it can still be seen as an exhibit.[42,43]

Further reading

- Hubbell, J. G. & Smith, R. W. – **Neptune in America –Negotiating a Discovery** – History of Astronomy Journal V.23, NO. 4/NOV, P.261, 1992 (Bibliographic Code: 1992JHA....23..261H)[44]
- Lecture notes with orbital positions at the time of discovery.[45]

Bibliography

- Airy, G. B. (1847). "Account of some circumstances historically connected with the discovery of the planet exterior to Uranus". *Memoirs of the Royal Astronomical Society*. **16**: 385–414. Bibcode: 1847MmRAS..16..385A[46].
- Airy, W. (ed.) (1896). *The Autobiography of Sir George Biddell Airy*[47]. Cambridge University Press. from Project Gutenberg
- [Anon.] (2001) "Bouvard, Alexis", *Encyclopædia Britannica*, Deluxe CDROM edition
- Baum, R.; Sheehan, W. (1997). *In Search of Planet Vulcan: The Ghost in Newton's Clockwork Universe*. Plenum. ISBN 0-306-45567-6.
- Bouvard, A. (1821), *Tables astronomiques publiées par le Bureau des Longitudes de France*[48], Paris, FR: Bachelier
- Chapman, A. (1988). "Private research and public duty: George Biddell Airy and the search for Neptune". *Journal for the History of Astronomy*. **19** (2): 121–139. Bibcode: 1988JHA....19..121C[49].
- Dieke, S. (1970). "Heinrich Louis D' Arrest". *Dictionary of Scientific Biography*. **1**. New York: Charles Scribner's Sons. pp. 295–296. ISBN 0-684-10114-9.
- Doggett, L. E. (1997) "Celestial mechanics", in Lankford, J. (ed.) (1997). *History of Astronomy, an Encyclopedia*. pp. 131–40.
- Dreyer, J. L. E. & Turner, H. H. (eds) (1987) [1923]. *History of the Royal Astronomical Society* [*1*]: *1820–1920*. pp. 161–2.

- Grosser, M. (1962). *The Discovery of Neptune*. Harvard University Press. ISBN 0-674-21225-8.
- — (1970). "Adams, John Couch". *Dictionary of Scientific Biography*. **1**. New York: Charles Scribner's Sons. pp. 53–54. ISBN 0-684-10114-9.
- Harrison, H. M (1994). *Voyager in Time and Space: The Life of John Couch Adams, Cambridge Astronomer*. Lewes: Book Guild, ISBN 0-86332-918-7
- Hughes, D. W. (1996). "J. C. Adams, Cambridge and Neptune". *Notes and Records of the Royal Society*. **50** (2): 245–8. doi: 10.1098/rsnr.1996.0027[50].
- Hutchins, R. (2004) " Adams, John Couch (1819–1892)[51]", *Oxford Dictionary of National Biography*, Oxford University Press, accessed 23 August 2007 (subscription or UK public library membership[52] required)
- J. W. L. G. [J. W. L. Glaisher] (March 1882). "James Challis". *Monthly Notices of the Royal Astronomical Society*. **43**: 160–79. Bibcode: 1883MNRAS..43..160.[53]. doi: 10.1093/mnras/43.4.160[54].
- Kollerstrom, Nick (2001). "Neptune's Discovery. The British Case for Co-Prediction"[55]. Unuiversity College London. Archived from the original[56] on 2005-11-11. Retrieved 2007-03-19.
- Moore, P. (1996). *The Planet Neptune: An Historical Survey before Voyager*. Praxis. ISBN 0-471-96015-2.
- Nichol, J. P. (1855). *The Planet Neptune: An Exposition and History*. Edinburgh: James Nichol.
- O'Connor, J. J.; Robertson, E. F. (1996). "Mathematical discovery of planets"[57]. *MacTutor History of Mathematics archive*. University of St. Andrews. Retrieved 2007-09-17.
- Rawlins, Dennis (1992). "The Neptune Conspiracy"[58] (PDF). *DIO, the International Journal of Scientific History*. **2** (3): 115–142.
- Rawlins, Dennis (1994). "Theft of the Neptune papers"[59] (PDF). *DIO, the International Journal of Scientific History*. **4** (2): 92–102. Bibcode: 1994DIO.....4...92R[60].
- Rawlins, Dennis (1999). "British Neptune Disaster File Recovered"[61] (PDF). *DIO, the International Journal of Scientific History*. **9** (1): 3–25.
- Sampson, R.A. (1904). "A description of Adams's manuscripts on the perturbations of Uranus". *Memoirs of the Royal Astronomical Society*. **54**: 143–161. Bibcode: 1904MmRAS..54..143S[62].
- Sheehan, W.; Baum, R. (September 1996). "Neptune's Discovery 150 Years Later". *Astronomy*: 42–49.
- Sheehan, W.; Thurber, S. (2007). "John Couch Adams's Asperger syndrome and the British non-discovery of Neptune". *Notes and Records of the Royal Society*. **61** (3): 285–299. doi: 10.1098/rsnr.2007.0187[63].

- Sheehan, W. *et al.* (2004). The Case of the Pilfered Planet — Did the British steal Neptune?[64] *Scientific American*
- Smart, W. M. (1946). "John Couch Adams and the discovery of Neptune". *Nature*. **158** (4019): 648–652. Bibcode: 1946Natur.158..648S[65]. doi: 10.1038/158648a0[66].
- — (1947). "John Couch Adams and the discovery of Neptune". *Occasional Notes of the Royal Astronomical Society*. **2**: 33–88.
- Smith, R. W. (1989). "The Cambridge network in action: the discovery of Neptune". *Isis*. **80** (303): 395–422. doi: 10.1086/355082[67].
- Standage, T. (2000). *The Neptune File*. Penguin Press.
- Unknown. (October 11, 1980). "Did Galileo See Neptune?". *Science News*. **118** (15): 231. doi: 10.2307/3965133[68]. JSTOR 3965133[69].

Physical characteristics

Extraterrestrial diamonds

Although diamonds on Earth are relatively rare, **extraterrestrial diamonds** (diamonds formed outside of Earth) are very common. Microscopic diamonds not much larger than molecules are abundant in meteorites and some of them retain a record of their formation in stars before the Solar System existed. High pressure experiments suggest large quantities of diamonds are formed from methane on the ice giant planets Uranus and Neptune, while some extrasolar planets may be composed almost entirely of diamond. Diamonds are also found in stars and may have been the first mineral ever to have formed.

Meteorites

In 1987, a team of scientists examined some meteorites and found grains of diamond about 2 nanometers in diameter (nanodiamonds). Trapped in them were noble gases whose isotopic signature indicated they came from outside the Solar System. Analyses of more meteorites found nanodiamonds from many different stars. The record of their origins was preserved despite a long and violent history that started when they were ejected from a star into the interstellar medium, went through the formation of the Solar System, were incorporated into a planetary body that was later broken up into meteorites, and finally crashed on the Earth's surface.

In meteorites, nanodiamonds make up about 3 percent of the carbon and 400 parts per million of the mass. Grains of silicon carbide and graphite also have anomalous isotopic patterns. Collectively they are known as *presolar grains* or *stardust* and their properties constrain models of nucleosynthesis in giant stars and supernovae.

It is unclear how many nanodiamonds in meteorites are really from outside the Solar System. Only a very small fraction of them contain noble gases and until

Figure 26: *Artist's conception of a multitude of tiny diamonds next to a hot star.*

recently it was not possible to study them individually. On average, the ratio of carbon-12 to carbon-13 matches that of the Earth's atmosphere while that of nitrogen-14 to nitrogen-15 matches the Sun. Techniques such as atom probe tomography will make it possible to examine individual grains, but due to the limited number of atoms, the isotopic resolution is limited.

If most nanodiamonds did form in the Solar System, that raises the question of how this is possible. On the surface of Earth, graphite is the stable carbon mineral while larger diamonds can only be formed in the kind of temperatures and pressures that are found deep in the mantle. However, nanodiamonds are close to molecular size: one with a diameter of 2.8 nm, the median size, contains about 1800 carbon atoms. In very small minerals, surface energy is important and diamonds are more stable than graphite because the diamond structure is more compact. The crossover in stability is between 1 and 5 nm. At even smaller sizes, a variety of other forms of carbon such as fullerenes can be found as well as diamond cores wrapped in fullerenes.

The most carbon-rich meteorites, with abundances up to 7 parts per thousand by weight, are ureilites.[241] These have no known parent body and their origin is controversial. Diamonds are common in highly shocked ureilites, and most are thought to have been formed by either the shock of the impact with Earth or with other bodies in space.[264] However, much larger diamonds were found in fragments of a meteorite called Almahata Sitta, found in the Nubian desert of Sudan. They contained inclusions of iron- and sulfur-bearing minerals, the first inclusions to be found in extraterrestrial diamonds. They were dated at 4.5 billion-year-old crystals and were formed at pressures greater than 20 gigapascals. The authors of a 2018 study concluded that they must have come from a protoplanet with a size between that of the moon and Mars but is no longer intact.

Infrared emissions from space, observed by the Infrared Space Observatory and the Spitzer Space Telescope, has made it clear that carbon-containing molecules are ubiquitous in space. These include polycyclic aromatic hydrocarbons (PAHs), fullerenes and diamondoids (hydrocarbons that have the same crystal structure as diamond). If dust in space has a similar concentration, a gram of it would carry up to 10 quadrillion of them, but so far there is little evidence for their presence in the interstellar medium; they are difficult to tell apart from diamondoids.

A 2014 study led by James Kennett at the University of California Santa Barbara identified a thin layer of diamonds spread over three continents. This lent support to a contentious hypothesis that a collision of a large comet with the Earth about 13,000 years ago caused the extinction of megafauna in North America and put an end to the Clovis culture.

Planets

Solar System

In 1981, Marvin Ross wrote a paper titled "The ice layer in Uranus and Neptune—diamonds in the sky?" in which he proposed that huge quantities of diamonds might be found in the interior of these planets. At Lawrence Livermore, he had analyzed data from shock-wave compression of methane (CH_4) and found that the extreme pressure separated the carbon atom from the hydrogen, freeing it to form diamond.

Theoretical modeling by Sandro Scandolo and others predicted that diamonds would form at pressures over 300 gigapascals (GPa), but even at lower pressures methane would be disrupted and form chains of hydrocarbons. High pressure experiments at the University of California Berkeley using a diamond anvil cell found both phenomena at only 50 GPa and a temperature of 2500 kelvins, equivalent to depths of 7000 kilometers below Neptune's cloud tops. Another experiment at the Geophysical Laboratory saw methane becoming unstable at only 7 GPa and 2000 kelvins. After forming, denser diamonds would sink. This "diamond rain" would convert potential energy into heat and help drive the convection that generates Neptune's magnetic field.

Figure 27: *Uranus, imaged by Voyager 2 in 1986.*

There are some uncertainties in how well the experimental results apply to Uranus and Neptune. Water and hydrogen mixed with the methane may alter the chemical reactions. A physicist at the Fritz Haber Institute in Berlin showed that the carbon on these planets is not concentrated enough to form diamonds from scratch. A proposal that diamonds may also form in Jupiter and Saturn, where the concentration of carbon is far lower, was considered unlikely because the diamonds would quickly dissolve.

Experiments looking for conversion of methane to diamonds found weak signals and did not reach the temperatures and pressures expected in Uranus and Neptune. However, a recent experiment used shock heating by lasers to reach temperatures and pressures expected at a depth of 10,000 kilometers below the surface of Uranus. When they did this to polystyrene, nearly every carbon atom in the material was incorporated into diamond crystals within a nanosecond.

Extrasolar

In the Solar System, 70% to 90% of the rocky planets (Venus, Earth and Mars) consist of silicates. By contrast, stars with a high ratio of carbon to oxygen may be orbited by planets that are mostly carbides, with the most common material being silicon carbide. This has a higher thermal conductivity and a lower

Figure 28: *On Earth, the natural form of silicon carbide is a rare mineral, moissanite.*

thermal expansivity than silicates. This would result in more rapid conductive cooling near the surface, but lower down the convection could be at least as vigorous as that in silicate planets.

One such planet is PSR J1719-1438 b, companion to a millisecond pulsar. It has a density at least twice that of lead, and may be composed mainly of ultra-dense diamond. It is believed to be the remnant of a white dwarf after the pulsar stripped away more than 99 percent of its mass.

Another planet, 55 Cancri e, has been called a "super-Earth" because, like Earth, it is a rocky planet orbiting a sun-like star, but it has twice the radius and eight times the mass. The researchers who discovered it in 2012 concluded that it was carbon-rich, making an abundance of diamond likely. However, later analyses using multiple measures for the star's chemical composition indicated that the star has 25 percent more oxygen than carbon. This makes it less likely that the planet itself is a carbon planet.

Stars

It has been proposed that diamonds exist in carbon-rich stars, particularly white dwarfs; and carbonado, a polycrystalline mix of diamond, graphite and amorphous carbon and the toughest natural form of carbon, could come from supernovae and white dwarfs. The largest white dwarf found in the universe so far, BPM 37093, located 50 light-years (4.7×10^{14} km) away in the constellation Centaurus and having a diameter of 2,500-mile (4,000 km), may have a diamond core.

In 2008, Robert Hazen and colleagues at the Carnegie Institution in Washington, D.C. published a paper, "Mineral evolution", in which they explored the history of mineral formation and found that the diversity of minerals has changed over time as the conditions have changed. Before the Solar System formed, only a small number of minerals were present, including diamonds and olivine. The first minerals may have been small diamonds formed in stars because stars are rich in carbon and diamonds form at a higher temperature than any other known mineral.

Orbital resonances

Kuiper belt

Types of distant minor planets

- Cis-Neptunian objects
  - Centaurs
  - Neptune trojans
- Trans-Neptunian objects (TNOs)‡
  - Kuiper belt objects (KBOs)
    - Classical KBOs (cubewanos)
    - Resonant KBOs
      - Plutinos (2:3 resonance)
  - Scattered disc objects (SDOs)
    - Resonant SDOs
  - Detached objects
  - Sednoids
  - Oort cloud objects (ICO/OCOs)

‡ Trans-Neptunian dwarf planets are called "plutoids"

- $\underline{v}$
- $\underline{t}$
- $\underline{e}$[71]

The **Kuiper belt** (/'kaɪpər/),[72] occasionally called the **Edgeworth–Kuiper belt**, is a circumstellar disc in the outer Solar System, extending from the orbit of Neptune (at 30 AU) to approximately 50 AU from the Sun. It is similar to the asteroid belt, but is far larger—20 times as wide and 20 to 200 times as massive. Like the asteroid belt, it consists mainly of small bodies or remnants from when the Solar System formed. While many asteroids are composed primarily of rock and metal, most Kuiper belt objects are composed largely of frozen volatiles (termed "ices"), such as methane, ammonia and water. The Kuiper belt is home to three officially recognized dwarf planets: Pluto, Haumea and Makemake. Some of the Solar System's moons, such as Neptune's Triton and Saturn's Phoebe, may have originated in the region.[73]

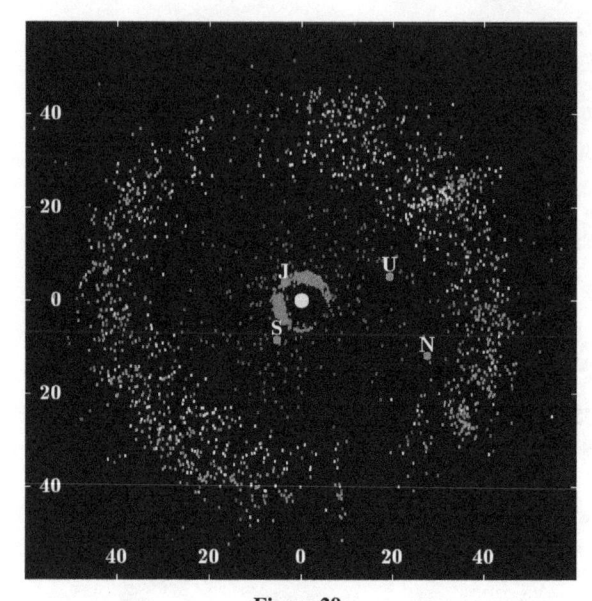

Figure 29:
Known objects in the Kuiper belt beyond the orbit of Neptune. (Scale in AU; epoch as of January 2015.)
Distances but not sizes are to scale
Source: Minor Planet Center, www<wbr/>.cfeps<wbr/>.net[70] and others

The Kuiper belt was named after Dutch-American astronomer Gerard Kuiper, though he did not predict its existence. In 1992, Albion was discovered, the first Kuiper belt object (KBO) since Pluto and Charon. Since its discovery, the number of known KBOs has increased to over a thousand, and more than 100,000 KBOs over 100 km (62 mi) in diameter are thought to exist.[74] The Kuiper belt was initially thought to be the main repository for periodic comets, those with orbits lasting less than 200 years. Studies since the mid-1990s have shown that the belt is dynamically stable and that comets' true place of origin is the scattered disc, a dynamically active zone created by the outward motion of Neptune 4.5 billion years ago; scattered disc objects such as Eris have extremely eccentric orbits that take them as far as 100 AU from the Sun.[75] Because the International Astronomical Union's Minor Planet Center, the body responsible for cataloguing minor planets in the Solar System, makes the distinction, the editorial choice for Wikipedia articles on the trans-Neptunian region is to make this distinction as well. On Wikipedia, Eris, the most-massive known trans-Neptunian object, is not part of the Kuiper belt and this makes Pluto the most-massive Kuiper belt object.</ref>

The Kuiper belt is distinct from the theoretical Oort cloud, which is a thousand times more distant and is mostly spherical. The objects within the Kuiper belt, together with the members of the scattered disc and any potential Hills cloud or Oort cloud objects, are collectively referred to as trans-Neptunian objects (TNOs). Pluto is the largest and most massive member of the Kuiper belt, and the largest and the second-most-massive known TNO, surpassed only by Eris in the scattered disc. Originally considered a planet, Pluto's status as part of the Kuiper belt caused it to be reclassified as a dwarf planet in 2006. It is compositionally similar to many other objects of the Kuiper belt and its orbital period is characteristic of a class of KBOs, known as "plutinos", that share the same 2:3 resonance with Neptune.

History

After the discovery of Pluto in 1930, many speculated that it might not be alone. The region now called the Kuiper belt was hypothesized in various forms for decades. It was only in 1992 that the first direct evidence for its existence was found. The number and variety of prior speculations on the nature of the Kuiper belt have led to continued uncertainty as to who deserves credit for first proposing it.[76]

Hypotheses

The first astronomer to suggest the existence of a trans-Neptunian population was Frederick C. Leonard. Soon after Pluto's discovery by Clyde Tombaugh in 1930, Leonard pondered whether it was "not likely that in Pluto there has come to light the *first* of a *series* of ultra-Neptunian bodies, the remaining members of which still await discovery but which are destined eventually to be detected". That same year, astronomer Armin O. Leuschner suggested that Pluto "may be one of many long-period planetary objects yet to be discovered."

In 1943, in the *Journal of the British Astronomical Association*, Kenneth Edgeworth hypothesized that, in the region beyond Neptune, the material within the primordial solar nebula was too widely spaced to condense into planets, and so rather condensed into a myriad of smaller bodies. From this he concluded that "the outer region of the solar system, beyond the orbits of the planets, is occupied by a very large number of comparatively small bodies" and that, from time to time, one of their number "wanders from its own sphere and appears as an occasional visitor to the inner solar system",[77] becoming a comet.

In 1951, in a paper in *Astrophysics: A Topical Symposium*, Gerard Kuiper speculated on a similar disc having formed early in the Solar System's evolution, but he did not think that such a belt still existed today. Kuiper was

Figure 30: *Astronomer Gerard Kuiper, after whom the Kuiper belt is named*

operating on the assumption, common in his time, that Pluto was the size of Earth and had therefore scattered these bodies out toward the Oort cloud or out of the Solar System. Were Kuiper's hypothesis correct, there would not be a Kuiper belt today.

The hypothesis took many other forms in the following decades. In 1962, physicist Al G.W. Cameron postulated the existence of "a tremendous mass of small material on the outskirts of the solar system".[78] In 1964, Fred Whipple, who popularised the famous "dirty snowball" hypothesis for cometary structure, thought that a "comet belt" might be massive enough to cause the purported discrepancies in the orbit of Uranus that had sparked the search for Planet X, or, at the very least, massive enough to affect the orbits of known comets. Observation ruled out this hypothesis.

In 1977, Charles Kowal discovered 2060 Chiron, an icy planetoid with an orbit between Saturn and Uranus. He used a blink comparator, the same device that had allowed Clyde Tombaugh to discover Pluto nearly 50 years before. In 1992, another object, 5145 Pholus, was discovered in a similar orbit. Today, an entire population of comet-like bodies, called the centaurs, is known to exist in the region between Jupiter and Neptune. The centaurs' orbits are unstable and have dynamical lifetimes of a few million years. From the time of Chiron's discovery in 1977, astronomers have speculated that the centaurs therefore must be frequently replenished by some outer reservoir.[79]

Further evidence for the existence of the Kuiper belt later emerged from the study of comets. That comets have finite lifespans has been known for some time. As they approach the Sun, its heat causes their volatile surfaces to sublimate into space, gradually dispersing them. In order for comets to continue to be visible over the age of the Solar System, they must be replenished frequently. One such area of replenishment is the Oort cloud, a spherical swarm of comets extending beyond 50,000 AU from the Sun first hypothesised by Dutch astronomer Jan Oort in 1950. The Oort cloud is thought to be the point of origin of long-period comets, which are those, like Hale–Bopp, with orbits lasting thousands of years.[80]

There is another comet population, known as short-period or periodic comets, consisting of those comets that, like Halley's Comet, have orbital periods of less than 200 years. By the 1970s, the rate at which short-period comets were being discovered was becoming increasingly inconsistent with their having emerged solely from the Oort cloud.[81] For an Oort cloud object to become a short-period comet, it would first have to be captured by the giant planets. In a paper published in *Monthly Notices of the Royal Astronomical Society* in 1980, Uruguayan astronomer Julio Fernández stated that for every short-period comet to be sent into the inner Solar System from the Oort cloud, 600 would have to be ejected into interstellar space. He speculated that a comet belt from between 35 and 50 AU would be required to account for the observed number of comets. Following up on Fernández's work, in 1988 the Canadian team of Martin Duncan, Tom Quinn and Scott Tremaine ran a number of computer simulations to determine if all observed comets could have arrived from the Oort cloud. They found that the Oort cloud could not account for all short-period comets, particularly as short-period comets are clustered near the plane of the Solar System, whereas Oort-cloud comets tend to arrive from any point in the sky. With a "belt", as Fernández described it, added to the formulations, the simulations matched observations. Reportedly because the words "Kuiper" and "comet belt" appeared in the opening sentence of Fernández's paper, Tremaine named this hypothetical region the "Kuiper belt".[82]

Discovery

In 1987, astronomer David Jewitt, then at MIT, became increasingly puzzled by "the apparent emptiness of the outer Solar System". He encouraged then-graduate student Jane Luu to aid him in his endeavour to locate another object beyond Pluto's orbit, because, as he told her, "If we don't, nobody will."[83] Using telescopes at the Kitt Peak National Observatory in Arizona and the Cerro Tololo Inter-American Observatory in Chile, Jewitt and Luu conducted their search in much the same way as Clyde Tombaugh and Charles Kowal had, with a blink comparator. Initially, examination of each pair of plates took

Figure 31: *The array of telescopes atop Mauna Kea, with which the Kuiper belt was discovered*

about eight hours,[84] but the process was sped up with the arrival of electronic charge-coupled devices or CCDs, which, though their field of view was narrower, were not only more efficient at collecting light (they retained 90% of the light that hit them, rather than the 10% achieved by photographs) but allowed the blinking process to be done virtually, on a computer screen. Today, CCDs form the basis for most astronomical detectors.[85] In 1988, Jewitt moved to the Institute of Astronomy at the University of Hawaii. Luu later joined him to work at the University of Hawaii's 2.24 m telescope at Mauna Kea.[86] Eventually, the field of view for CCDs had increased to 1024 by 1024 pixels, which allowed searches to be conducted far more rapidly.[87] Finally, after five years of searching, Jewitt and Luu announced on August 30, 1992 the "Discovery of the candidate Kuiper belt object" 15760 Albion. Six months later, they discovered a second object in the region, (181708) 1993 FW.

Studies conducted since the trans-Neptunian region was first charted have shown that the region now called the Kuiper belt is not the point of origin of short-period comets, but that they instead derive from a linked population called the scattered disc. The scattered disc was created when Neptune migrated outward into the proto-Kuiper belt, which at the time was much closer to the Sun, and left in its wake a population of dynamically stable objects that could never be affected by its orbit (the Kuiper belt proper), and a population whose perihelia are close enough that Neptune can still disturb them as it travels around the Sun (the scattered disc). Because the scattered disc is dynamically active and the Kuiper belt relatively dynamically stable, the scattered disc is now seen as the most likely point of origin for periodic comets.

Figure 32: *Dust in the Kuiper belt creates a faint infrared disc. (Click on the "play" button to watch the video.)*

Name

Astronomers sometimes use the alternative name Edgeworth–Kuiper belt to credit Edgeworth, and KBOs are occasionally referred to as EKOs. Brian G. Marsden claims that neither deserves true credit: "Neither Edgeworth nor Kuiper wrote about anything remotely like what we are now seeing, but Fred Whipple did".[88] David Jewitt comments: "If anything... Fernández most nearly deserves the credit for predicting the Kuiper Belt."

KBOs are sometimes called "kuiperoids", a name suggested by Clyde Tombaugh.[89] The term "trans-Neptunian object" (TNO) is recommended for objects in the belt by several scientific groups because the term is less controversial than all others—it is not an exact synonym though, as TNOs include all objects orbiting the Sun past the orbit of Neptune, not just those in the Kuiper belt.

Structure

At its fullest extent (but excluding the scattered disc), including its outlying regions, the Kuiper belt stretches from roughly 30 to 55 AU. The main body of the belt is generally accepted to extend from the 2:3 mean-motion resonance (see below) at 39.5 AU to the 1:2 resonance at roughly 48 AU. The Kuiper belt is quite thick, with the main concentration extending as much as ten degrees outside the ecliptic plane and a more diffuse distribution of objects extending several times farther. Overall it more resembles a torus or doughnut than a belt. Its mean position is inclined to the ecliptic by 1.86 degrees.

The presence of Neptune has a profound effect on the Kuiper belt's structure due to orbital resonances. Over a timescale comparable to the age of the Solar System, Neptune's gravity destabilises the orbits of any objects that happen to lie in certain regions, and either sends them into the inner Solar System or out into the scattered disc or interstellar space. This causes the Kuiper belt to have pronounced gaps in its current layout, similar to the Kirkwood gaps in the asteroid belt. In the region between 40 and 42 AU, for instance, no objects can retain a stable orbit over such times, and any observed in that region must have migrated there relatively recently.

Classical belt

Between the 2:3 and 1:2 resonances with Neptune, at approximately 42–48 AU, the gravitational interactions with Neptune occur over an extended timescale, and objects can exist with their orbits essentially unaltered. This region is known as the classical Kuiper belt, and its members comprise roughly two thirds of KBOs observed to date. Because the first modern KBO discovered, (15760) 1992 QB1, is considered the prototype of this group, classical KBOs are often referred to as cubewanos ("Q-B-1-os"). The guidelines established by the IAU demand that classical KBOs be given names of mythological beings associated with creation.

The classical Kuiper belt appears to be a composite of two separate populations. The first, known as the "dynamically cold" population, has orbits much like the planets; nearly circular, with an orbital eccentricity of less than 0.1, and with relatively low inclinations up to about 10° (they lie close to the plane of the Solar System rather than at an angle). The cold population also contain a concentration of objects, referred to as the kernel, with semi-major axes at 44–44.5 AU. The second, the "dynamically hot" population, has orbits much more inclined to the ecliptic, by up to 30°. The two populations have been named this way not because of any major difference in temperature, but from analogy to particles in a gas, which increase their relative velocity as they become heated up. Not only are the two populations in different orbits, the cold population also differs in color and albedo, being redder and brighter, has a larger fraction of binary objects, has a different size distribution, and lacks very large objects. The difference in colors may be a reflection of different compositions, which suggests they formed in different regions. The hot population is proposed to have formed near Neptune's original orbit and to have been scattered out during the migration of the giant planets. The cold population, on the other hand, has been proposed to have formed more or less in its current position because the loose binaries would be unlikely to survive

Kuiper belt

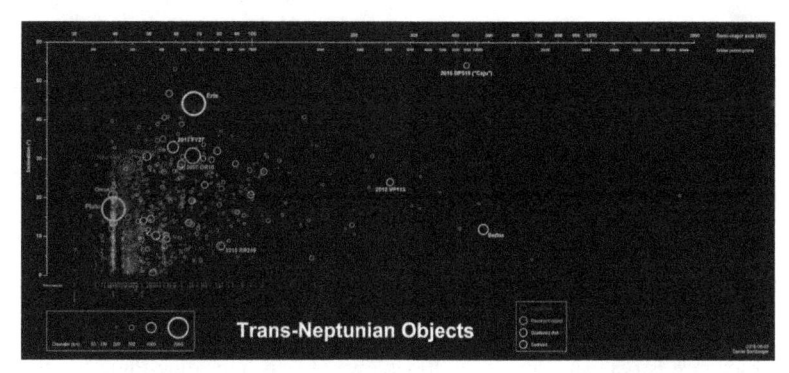

Figure 33: *Distribution of cubewanos (blue), Resonant trans-Neptunian objects (red), Sednoids (yellow) and scattered objects (grey)*

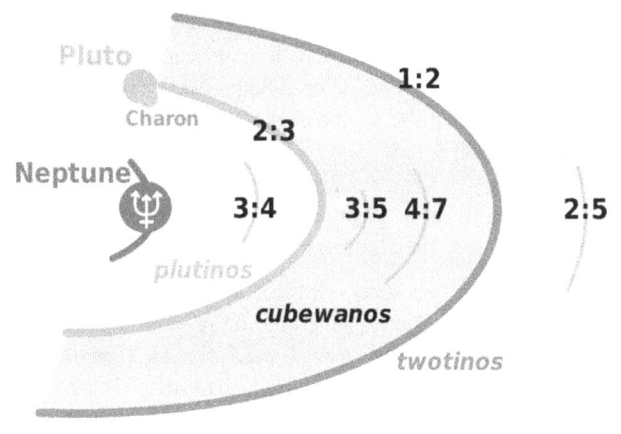

Figure 34: *Orbit classification (schematic of semi-major axes)*

encounters with Neptune. Although the Nice model appears to be able to at least partially explain a compositional difference, it has also been suggested the color difference may reflect differences in surface evolution.

Resonances

When an object's orbital period is an exact ratio of Neptune's (a situation called a mean-motion resonance), then it can become locked in a synchronised motion with Neptune and avoid being perturbed away if their relative alignments are appropriate. If, for instance, an object orbits the Sun twice for every three Neptune orbits, and if it reaches perihelion with Neptune a quarter of an orbit away from it, then whenever it returns to perihelion, Neptune will always be in about the same relative position as it began, because it will have completed 1 $1/2$ orbits in the same time. This is known as the 2:3 (or 3:2) resonance, and it corresponds to a characteristic semi-major axis of about 39.4 AU. This 2:3 resonance is populated by about 200 known objects, including Pluto together with its moons. In recognition of this, the members of this family are known as plutinos. Many plutinos, including Pluto, have orbits that cross that of Neptune, though their resonance means they can never collide. Plutinos have high orbital eccentricities, suggesting that they are not native to their current positions but were instead thrown haphazardly into their orbits by the migrating Neptune. IAU guidelines dictate that all plutinos must, like Pluto, be named for underworld deities. The 1:2 resonance (whose objects complete half an orbit for each of Neptune's) corresponds to semi-major axes of ∼47.7 AU, and is sparsely populated. Its residents are sometimes referred to as twotinos. Other resonances also exist at 3:4, 3:5, 4:7 and 2:5.[90] Neptune has a number of trojan objects, which occupy its Lagrangian points, gravitationally stable regions leading and trailing it in its orbit. Neptune trojans are in a 1:1 mean-motion resonance with Neptune and often have very stable orbits.

Additionally, there is a relative absence of objects with semi-major axes below 39 AU that cannot apparently be explained by the present resonances. The currently accepted hypothesis for the cause of this is that as Neptune migrated outward, unstable orbital resonances moved gradually through this region, and thus any objects within it were swept up, or gravitationally ejected from it.[91]

Kuiper cliff

The 1:2 resonance appears to be an edge beyond which few objects are known. It is not clear whether it is actually the outer edge of the classical belt or just the beginning of a broad gap. Objects have been detected at the 2:5 resonance at roughly 55 AU, well outside the classical belt; predictions of a large number of bodies in classical orbits between these resonances have not been verified through observation.

Based on estimations of the primordial mass required to form Uranus and Neptune, as well as bodies as large as Pluto (see below), earlier models of the Kuiper belt had suggested that the number of large objects would increase by a

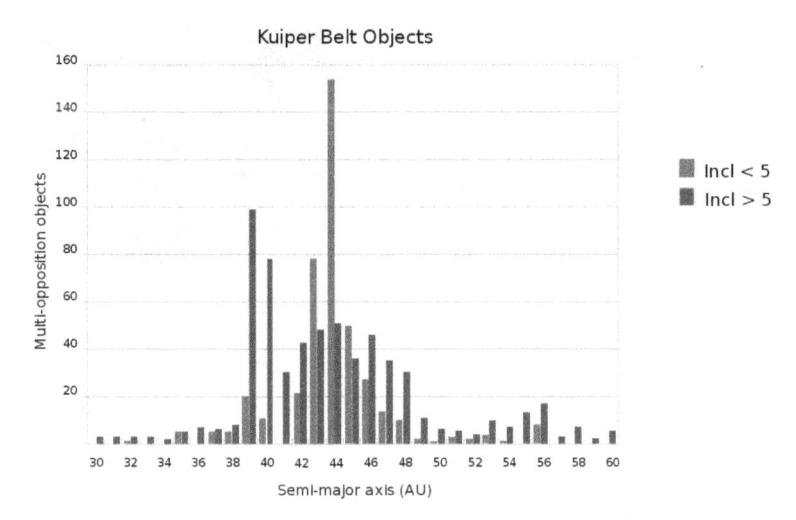

Figure 35: *Histogram of the semi-major axes of Kuiper belt objects with inclinations above and below 5 degrees. Spikes from the plutinos and the 'kernel' are visible at 39–40 AU and 44 AU.*

factor of two beyond 50 AU, so this sudden drastic falloff, known as the *Kuiper cliff*, was unexpected, and to date its cause is unknown. In 2003, Bernstein, Trilling, et al. found evidence that the rapid decline in objects of 100 km or more in radius beyond 50 AU is real, and not due to observational bias. Possible explanations include that material at that distance was too scarce or too scattered to accrete into large objects, or that subsequent processes removed or destroyed those that did. Patryk Lykawka of Kobe University claimed that the gravitational attraction of an unseen large planetary object, perhaps the size of Earth or Mars, might be responsible.

Origin

The precise origins of the Kuiper belt and its complex structure are still unclear, and astronomers are awaiting the completion of several wide-field survey telescopes such as Pan-STARRS and the future LSST, which should reveal many currently unknown KBOs. These surveys will provide data that will help determine answers to these questions.

The Kuiper belt is thought to consist of planetesimals, fragments from the original protoplanetary disc around the Sun that failed to fully coalesce into planets and instead formed into smaller bodies, the largest less than 3,000 kilometres

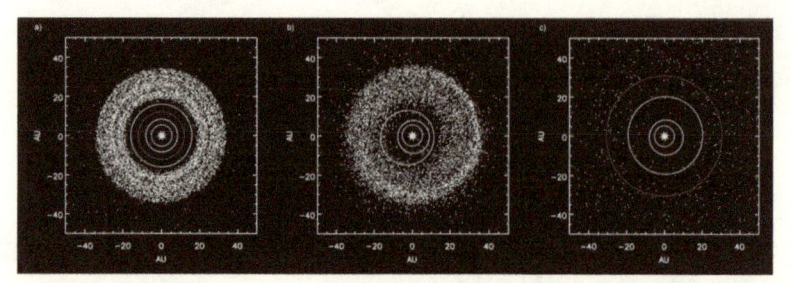

Figure 36: *Simulation showing outer planets and Kuiper belt: a) before Jupiter/Saturn 1:2 resonance, b) scattering of Kuiper belt objects into the Solar System after the orbital shift of Neptune, c) after ejection of Kuiper belt bodies by Jupiter*

(1,900 mi) in diameter. Studies of the crater counts on Pluto and Charon revealed a scarcity of small craters suggesting that such objects formed directly as sizeable objects in the range of tens of kilometers in diameter rather than being accreted from much smaller, roughly kilometer scale bodies. Hypothetical mechanisms for the formation of these larger bodies include the gravitational collapse of clouds of pebbles concentrated between eddies in a turbulent protoplanetary disk or in streaming instabilities. These collapsing clouds may fragment, forming binaries.

Modern computer simulations show the Kuiper belt to have been strongly influenced by Jupiter and Neptune, and also suggest that neither Uranus nor Neptune could have formed in their present positions, because too little primordial matter existed at that range to produce objects of such high mass. Instead, these planets are estimated to have formed closer to Jupiter. Scattering of planetesimals early in the Solar System's history would have led to migration of the orbits of the giant planets: Saturn, Uranus, and Neptune drifted outwards, whereas Jupiter drifted inwards. Eventually, the orbits shifted to the point where Jupiter and Saturn reached an exact 1:2 resonance; Jupiter orbited the Sun twice for every one Saturn orbit. The gravitational repercussions of such a resonance ultimately destabilized the orbits of Uranus and Neptune, causing them to be scattered outward onto high-eccentricity orbits that crossed the primordial planetesimal disc. While Neptune's orbit was highly eccentric, its mean-motion resonances overlapped and the orbits of the planetesimals evolved chaotically, allowing planetesimals to wander outward as far as Neptune's 1:2 resonance to form a dynamically cold belt of low-inclination objects. Later, after its eccentricity decreased, Neptune's orbit expanded outward toward its current position. Many planetesimals were captured into and remain in resonances during this migration, others evolved onto higher-inclination and lower-eccentricity orbits and escaped from the resonances onto stable orbits.

Many more planetesimals were scattered inward, with small fractions being captured as Jupiter trojans, as irregular satellites orbiting the giant planets, and as outer belt asteroids. The remainder were scattered outward again by Jupiter and in most cases ejected from the Solar System reducing the primordial Kuiper belt population by 99% or more.

The original version of the currently most popular model, the "Nice model", reproduces many characteristics of the Kuiper belt such as the "cold" and "hot" populations, resonant objects, and a scattered disc, but it still fails to account for some of the characteristics of their distributions. The model predicts a higher average eccentricity in classical KBO orbits than is observed (0.10–0.13 versus 0.07) and its predicted inclination distribution contains too few high inclination objects. In addition, the frequency of binary objects in the cold belt, many of which are far apart and loosely bound, also poses a problem for the model. These are predicted to have been separated during encounters with Neptune, leading some to propose that the cold disc formed at its current location, representing the only truly local population of small bodies in the solar system.

A recent modification of the Nice model has the Solar System begin with five giant planets, including an additional ice giant, in a chain of mean-motion resonances. About 400 million years after the formation of the Solar System the resonance chain is broken. Instead of being scattered into the disc, the ice giants first migrate outward several AU. This divergent migration eventually leads to a resonance crossing, destabilizing the orbits of the planets. The extra ice giant encounters Saturn and is scattered inward onto a Jupiter-crossing orbit and after a series of encounters is ejected from the Solar System. The remaining planets then continue their migration until the planetesimal disc is nearly depleted with small fractions remaining in various locations.

As in the original Nice model, objects are captured into resonances with Neptune during its outward migration. Some remain in the resonances, others evolve onto higher-inclination, lower-eccentricity orbits, and are released onto stable orbits forming the dynamically hot classical belt. The hot belt's inclination distribution can be reproduced if Neptune migrated from 24 AU to 30 AU on a 30 Myr timescale. When Neptune migrates to 28 AU, it has a gravitational encounter with the extra ice giant. Objects captured from the cold belt into the 1:2 mean-motion resonance with Neptune are left behind as a local concentration at 44 AU when this encounter causes Neptune's semi-major axis to jump outward. The objects deposited in the cold belt include some loosely bound 'blue' binaries originating from closer than the cold belt's current location. If Neptune's eccentricity remains small during this encounter, the chaotic evolution of orbits of the original Nice model is avoided and a primordial cold belt is preserved. In the later phases of Neptune's migration, a slow sweeping

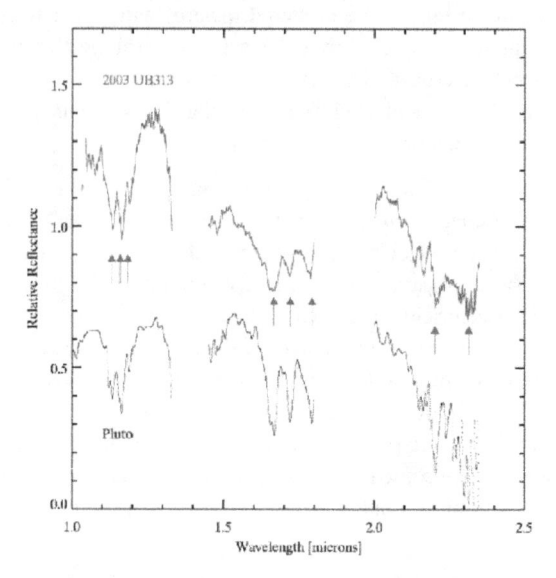

Figure 37: *The infrared spectra of both Eris and Pluto, highlighting their common methane absorption lines*

of mean-motion resonances removes the higher-eccentricity objects from the cold belt, truncating its eccentricity distribution.

Composition

Being distant from the Sun and major planets, Kuiper belt objects are thought to be relatively unaffected by the processes that have shaped and altered other Solar System objects; thus, determining their composition would provide substantial information on the makeup of the earliest Solar System. Due to their small size and extreme distance from Earth, the chemical makeup of KBOs is very difficult to determine. The principal method by which astronomers determine the composition of a celestial object is spectroscopy. When an object's light is broken into its component colors, an image akin to a rainbow is formed. This image is called a spectrum. Different substances absorb light at different wavelengths, and when the spectrum for a specific object is unravelled, dark lines (called absorption lines) appear where the substances within it have absorbed that particular wavelength of light. Every element or compound has its own unique spectroscopic signature, and by reading an object's full spectral "fingerprint", astronomers can determine its composition.

Figure 38: *Artist's impression of plutino and possible former C-type asteroid 2004 EW$_{95}$*

Analysis indicates that Kuiper belt objects are composed of a mixture of rock and a variety of ices such as water, methane, and ammonia. The temperature of the belt is only about 50 K, so many compounds that would be gaseous closer to the Sun remain solid. The densities and rock–ice fractions are known for only a small number of objects for which the diameters and the masses have been determined. The diameter can be determined by imaging with a high-resolution telescope such as the Hubble Space Telescope, by the timing of an occultation when an object passes in front of a star or, most commonly, by using the albedo of an object calculated from its infrared emissions. The masses are determined using the semi-major axes and periods of satellites, which are therefore known only for a few binary objects. The densities range from less than 0.4 to 2.6 g/cm^3. The least dense objects are thought to be largely composed of ice and have significant porosity. The densest objects are likely composed of rock with a thin crust of ice. There is a trend of low densities for small objects and high densities for the largest objects. One possible explanation for this trend is that ice was lost from the surface layers when differentiated objects collided to form the largest objects.

Initially, detailed analysis of KBOs was impossible, and so astronomers were only able to determine the most basic facts about their makeup, primarily their color. These first data showed a broad range of colors among KBOs, ranging from neutral grey to deep red. This suggested that their surfaces were composed of a wide range of compounds, from dirty ices to hydrocarbons. This diversity was startling, as astronomers had expected KBOs to be uniformly dark, having lost most of the volatile ices from their surfaces to the effects of cosmic rays.[92] Various solutions were suggested for this discrepancy, including resurfacing by impacts or outgassing. Jewitt and Luu's spectral analysis of the known Kuiper belt objects in 2001 found that the variation in color was too extreme to be easily explained by random impacts. The radiation from the Sun is thought to have chemically altered methane on the surface of KBOs, producing products such as tholins. Makemake has been shown to possess a number of hydrocarbons derived from the radiation-processing of methane, including ethane, ethylene and acetylene.

Although to date most KBOs still appear spectrally featureless due to their faintness, there have been a number of successes in determining their composition. In 1996, Robert H. Brown et al. acquired spectroscopic data on the KBO 1993 SC, which revealed that its surface composition is markedly similar to that of Pluto, as well as Neptune's moon Triton, with large amounts of methane ice. For the smaller objects, only colors and in some cases the albedos have been determined. These objects largely fall into two classes: gray with low albedos, or very red with higher albedos. The difference in colors and albedos is hypothesized to be due to the retention or the loss of hydrogen sulfide (H_2S) on the surface of these objects, with the surfaces of those that formed far enough from the Sun to retain H_2S being reddened due to irradiation.

The largest KBOs, such as Pluto and Quaoar, have surfaces rich in volatile compounds such as methane, nitrogen and carbon monoxide; the presence of these molecules is likely due to their moderate vapor pressure in the 30–50 K temperature range of the Kuiper belt. This allows them to occasionally boil off their surfaces and then fall again as snow, whereas compounds with higher boiling points would remain solid. The relative abundances of these three compounds in the largest KBOs is directly related to their surface gravity and ambient temperature, which determines which they can retain. Water ice has been detected in several KBOs, including members of the Haumea family such as 1996 TO_{66}, mid-sized objects such as 38628 Huya and 20000 Varuna, and also on some small objects. The presence of crystalline ice on large and mid-sized objects, including 50000 Quaoar where ammonia hydrate has also been detected, may indicate past tectonic activity aided by melting point lowering due to the presence of ammonia.

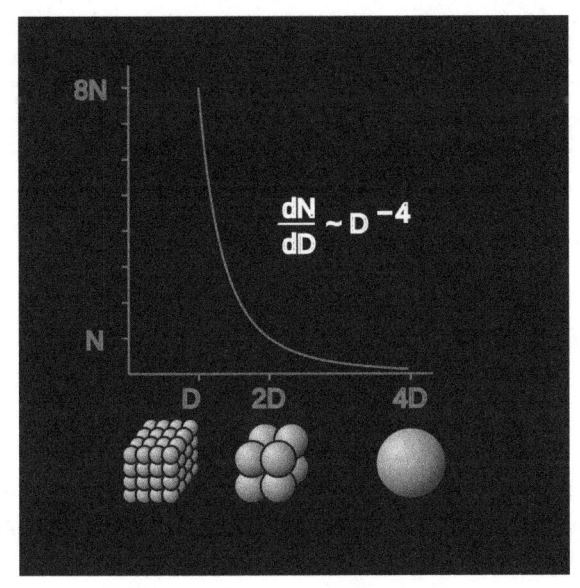

Figure 39: *Illustration of the power law*

Mass and size distribution

Despite its vast extent, the collective mass of the Kuiper belt is relatively low. The total mass is estimated to range between 1/25 and 1/10 the mass of the Earth. Conversely, models of the Solar System's formation predict a collective mass for the Kuiper belt of 30 Earth masses. This missing >99% of the mass can hardly be dismissed, because it is required for the accretion of any KBOs larger than 100 km (62 mi) in diameter. If the Kuiper belt had always had its current low density, these large objects simply could not have formed by the collision and mergers of smaller planetesimals. Moreover, the eccentricity and inclination of current orbits makes the encounters quite "violent" resulting in destruction rather than accretion. It appears that either the current residents of the Kuiper belt have been created closer to the Sun, or some mechanism dispersed the original mass. Neptune's current influence is too weak to explain such a massive "vacuuming", though the Nice model proposes that it could have been the cause of mass removal in the past. Although the question remains open, the conjectures vary from a passing star scenario to grinding of smaller objects, via collisions, into dust small enough to be affected by solar radiation. The extent of mass loss by collisional grinding is limited by the presence of loosely bound binaries in the cold disk, which are likely to be disrupted in collisions.

Bright objects are rare compared with the dominant dim population, as expected from accretion models of origin, given that only some objects of a given size would have grown further. This relationship between $N(D)$ (the number of objects of diameter greater than D) and D, referred to as brightness slope, has been confirmed by observations. The slope is inversely proportional to some power of the diameter D:

$$\frac{dN}{dD} \propto D^{-q}$$ where the current measures give q = 4 ±0.5.

This implies (assuming q is not 1) that

$$N \propto D^{1-q} + \text{a constant}.$$

(The constant may be non-zero only if the power law doesn't apply at high values of D.)

Less formally, if q is 4, for example, there are 8 (=2^3) times more objects in the 100–200 km range than in the 200–400 km range, and for every object with a diameter between 1000 and 1010 km there should be around 1000 (=10^3) objects with diameter of 100 to 101 km.

If q was 1 or less, the law would imply an infinite number and mass of large objects in the Kuiper belt. If $1 < q \leq 4$ there will be a finite number of objects greater than a given size, but the expected value of their combined mass would be infinite. If q is 4 or more, the law would imply an infinite mass of small objects. More accurate models find that the "slope" parameter q is in effect greater at large diameters and lesser at small diameters. It seems that Pluto is somewhat unexpectedly large, having several percent of the total mass of the Kuiper belt. It is not expected that anything larger than Pluto exists in the Kuiper belt, and in fact most of the brightest (largest) objects at inclinations less than 5° have probably been found.

For most TNOs, only the absolute magnitude is actually known, the size is inferred assuming a given albedo (not a safe assumption for larger objects).

Recent research has revealed that the size distributions of the hot classical and cold classical objects have differing slopes. The slope for the hot objects is q = 5.3 at large diameters and q = 2.0 at small diameters with the change in slope at 110 km. The slope for the cold objects is q = 8.2 at large diameters and q = 2.9 at small diameters with a change in slope at 140 km. The size distributions of the scattering objects, the plutinos, and the Neptune trojans have slopes similar to the other dynamically hot populations, but may instead have a divot, a sharp decrease in the number of objects below a specific size. This divot is hypothesized to be due to either the collisional evolution of the population, or to be due to the population having formed with no objects below this size, with the smaller objects being fragments of the original objects.

Kuiper belt

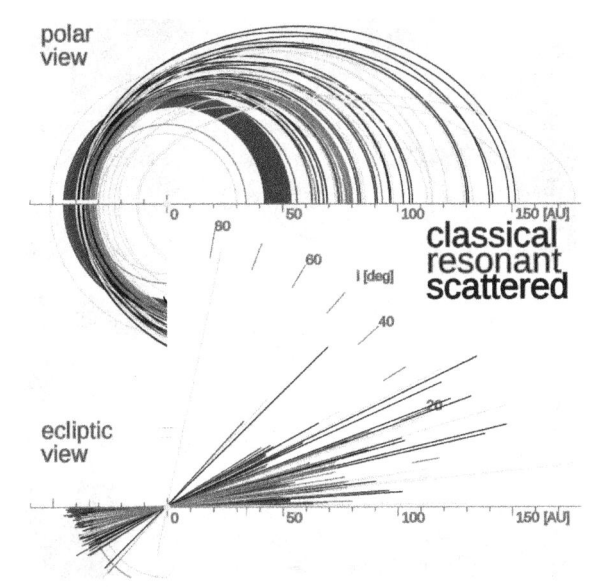

Figure 40: *Comparison of the orbits of scattered disc objects (black), classical KBOs (blue), and 2:5 resonant objects (green). Orbits of other KBOs are gray. (Orbital axes have been aligned for comparison.)*

As of December 2009, the smallest Kuiper belt object detected is 980 m across. It is too dim (magnitude 35) to be seen by *Hubble* directly, but it was detected by *Hubble*'s star tracking system when it occulted a star.

Scattered objects

The scattered disc is a sparsely populated region, overlapping with the Kuiper belt but extending to beyond 100 AU. Scattered disc objects (SDOs) have very elliptical orbits, often also very inclined to the ecliptic. Most models of Solar System formation show both KBOs and SDOs first forming in a primordial belt, with later gravitational interactions, particularly with Neptune, sending the objects outward, some into stable orbits (the KBOs) and some into unstable orbits, the scattered disc. Due to its unstable nature, the scattered disc is suspected to be the point of origin of many of the Solar System's short-period comets. Their dynamic orbits occasionally force them into the inner Solar System, first becoming centaurs, and then short-period comets.

Figure 41: *Neptune's moon Triton*

According to the Minor Planet Center, which officially catalogues all trans-Neptunian objects, a KBO, strictly speaking, is any object that orbits exclusively within the defined Kuiper belt region regardless of origin or composition. Objects found outside the belt are classed as scattered objects. In some scientific circles the term "Kuiper belt object" has become synonymous with any icy minor planet native to the outer Solar System assumed to have been part of that initial class, even if its orbit during the bulk of Solar System history has been beyond the Kuiper belt (e.g. in the scattered-disc region). They often describe scattered disc objects as "scattered Kuiper belt objects". Eris, which is known to be more massive than Pluto, is often referred to as a KBO, but is technically an SDO. A consensus among astronomers as to the precise definition of the Kuiper belt has yet to be reached, and this issue remains unresolved.

The centaurs, which are not normally considered part of the Kuiper belt, are also thought to be scattered objects, the only difference being that they were scattered inward, rather than outward. The Minor Planet Center groups the centaurs and the SDOs together as scattered objects.

Triton

During its period of migration, Neptune is thought to have captured a large KBO, Triton, which is the only large moon in the Solar System with a retrograde orbit (it orbits opposite to Neptune's rotation). This suggests that, unlike

Kuiper belt

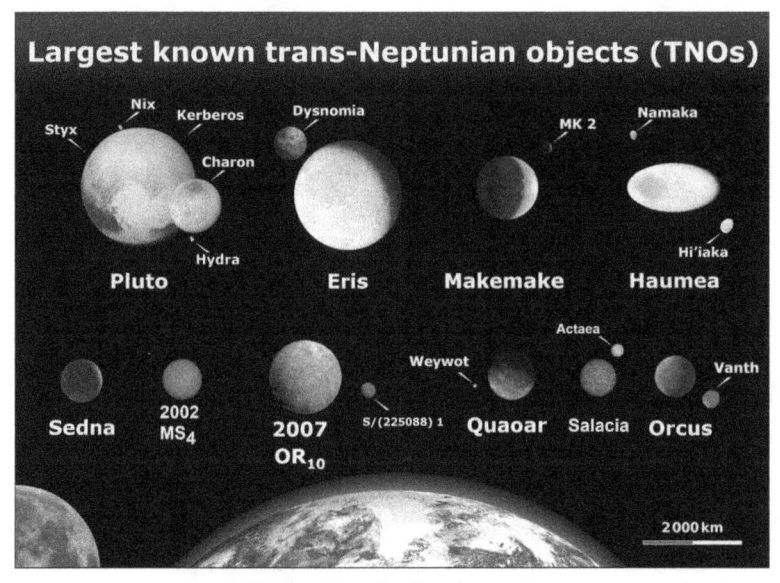

Figure 42: *style=float:bottom}}*

the large moons of Jupiter, Saturn and Uranus, which are thought to have coalesced from rotating discs of material around their young parent planets, Triton was a fully formed body that was captured from surrounding space. Gravitational capture of an object is not easy: it requires some mechanism to slow down the object enough to be caught by the larger object's gravity. A possible explanation is that Triton was part of a binary when it encountered Neptune. (Many KBOs are members of binaries. See below.) Ejection of the other member of the binary by Neptune could then explain Triton's capture. Triton is only 14% larger than Pluto, and spectral analysis of both worlds shows that their surfaces are largely composed of similar materials, such as methane and carbon monoxide. All this points to the conclusion that Triton was once a KBO that was captured by Neptune during its outward migration.

Largest KBOs

Since 2000, a number of KBOs with diameters of between 500 and 1,500 km (932 mi), more than half that of Pluto (diameter 2370 km), have been discovered. 50000 Quaoar, a classical KBO discovered in 2002, is over 1,200 km across. Makemake and Haumea, both announced on July 29, 2005, are larger still. Other objects, such as 28978 Ixion (discovered in 2001) and 20000 Varuna (discovered in 2000), measure roughly 500 km (311 mi) across.

Pluto

The discovery of these large KBOs in orbits similar to Pluto's led many to conclude that, aside from its relative size, Pluto was not particularly different from other members of the Kuiper belt. Not only are these objects similar to Pluto in size, but many also have satellites, and are of similar composition (methane and carbon monoxide have been found both on Pluto and on the largest KBOs). Thus, just as Ceres was considered a planet before the discovery of its fellow asteroids, some began to suggest that Pluto might also be reclassified.

The issue was brought to a head by the discovery of Eris, an object in the scattered disc far beyond the Kuiper belt, that is now known to be 27% more massive than Pluto. (Eris was originally thought to be larger than Pluto by volume, but the *New Horizons* mission found this not to be the case.) In response, the International Astronomical Union (IAU) was forced to define what a planet is for the first time, and in so doing included in their definition that a planet must have "cleared the neighbourhood around its orbit". As Pluto shares its orbit with many other sizable objects, it was deemed not to have cleared its orbit, and was thus reclassified from a planet to a dwarf planet, making it a member of the Kuiper belt.

Although Pluto is currently the largest known KBO, there is at least one known larger object currently outside the Kuiper belt that probably originated in it: Neptune's moon Triton (which, as explained above, is probably a captured KBO).

As of 2008, only five objects in the Solar System (Ceres, Eris, and the KBOs Pluto, Makemake and Haumea) are listed as dwarf planets by the IAU. 90482 Orcus, 28978 Ixion and many other Kuiper-belt objects are large enough to be in hydrostatic equilibrium; most of them will probably qualify when more is known about them.

Satellites

The six largest TNOs (Eris, Pluto, 2007 OR$_{10}$, Makemake, Haumea and Quaoar) are all known to have satellites, and two have more than one. A higher percentage of the larger KBOs have satellites than the smaller objects in the Kuiper belt, suggesting that a different formation mechanism was responsible. There are also a high number of binaries (two objects close enough in mass to be orbiting "each other") in the Kuiper belt. The most notable example is the Pluto–Charon binary, but it is estimated that around 11% of KBOs exist in binaries.

Figure 43: *Kuiper belt object—possible target of New Horizons spacecraft (artist's concept)*

Exploration

On January 19, 2006, the first spacecraft to explore the Kuiper belt, *New Horizons*, was launched, which flew by Pluto on July 14, 2015. Beyond the Pluto flyby, the mission's goal was to locate and investigate other, farther objects in the Kuiper belt.

On October 15, 2014, it was revealed that *Hubble* had uncovered three potential targets, provisionally designated PT1 ("potential target 1"), PT2 and PT3 by the *New Horizons* team. The objects' diameters were estimated to be in the 30–55 km range; too small to be seen by ground telescopes, at distances from the Sun of 43–44 AU, which would put the encounters in the 2018–2019 period. The initial estimated probabilities that these objects were reachable within *New Horizons*' fuel budget were 100%, 7%, and 97%, respectively. All were members of the "cold" (low-inclination, low-eccentricity) classical Kuiper belt, and thus very different from Pluto. PT1 (given the temporary designation "1110113Y" on the HST web site), the most favorably situated object, was magnitude 26.8, 30–45 km in diameter, and will be encountered around January 2019. Once sufficient orbital information was provided, the Minor Planet Center gave official designations to the three target KBOs: 2014 MU$_{69}$ (PT1), 2014 OS$_{393}$ (PT2), and 2014 PN$_{70}$ (PT3). By the fall of 2014,

Figure 44: *The KBO 2014 MU69 (green circles), the selected target for the New Horizons Kuiper belt object mission*

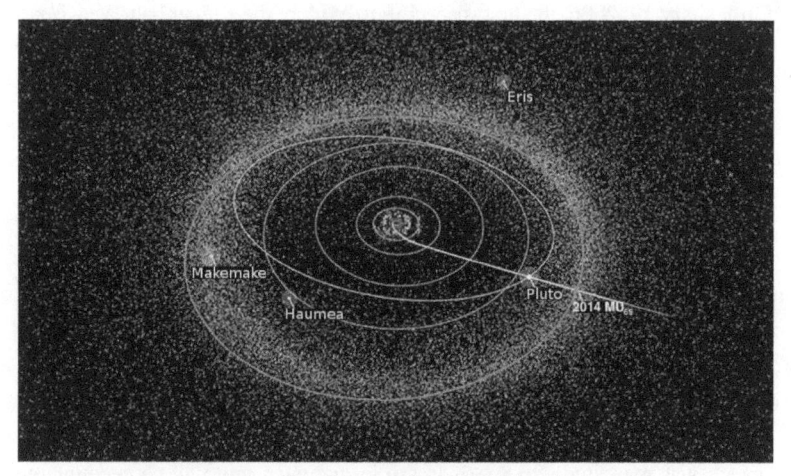

Figure 45: *Diagram showing the location of 2014 MU69 and trajectory for rendezvous*

a possible fourth target, 2014 MT69, had been eliminated by follow-up observations. PT2 was out of the running before the Pluto flyby.

On August 26, 2015, the first target, 2014 MU69, was chosen. Course adjustment took place in late October and early November 2015, leading to a flyby in January 2019. On July 1, 2016, NASA approved additional funding for *New Horizons* to visit the object.

On December 2, 2015, *New Horizons* detected 1994 JR1 from 270 million kilometres (170×10^6 mi) away, and the photographs show the shape of the object and one or two details.[93]

No follow up missions for *New Horizons* are planned, though at least two concepts for missions that would return to orbit or land on Pluto have been studied. Beyond Pluto, there exist many large KBOs that cannot be visited with *New Horizons*, such as the dwarf planets Makemake and Haumea. New missions would be tasked to explore and study these objects in detail. Thales Alenia Space has studied the logistics of an orbiter mission to Haumea, a high priority scientific target due to its status as the parent body of a collisional family that includes several other TNOs, as well as Haumea's ring and two moons. The lead author, Joel Poncy, has advocated for new technology that would allow spacecraft to reach and orbit KBOs in 10-20 years or less. *New Horizons* Principal Investigator Alan Stern has informally suggested missions that would flyby the planets Uranus or Neptune before visiting new KBO targets, thus furthering the exploration of the Kuiper belt while also visiting these ice giant planets for the first time since the *Voyager 2* flybys in the 1980s.

Quaoar would make a particularly attractive flyby target for a probe tasked with exploring the interstellar medium, as it currently lies near the heliospheric nose; Pontus Brandt at Johns Hopkins Applied Physics Laboratory and his colleagues have studied a probe that would flyby Quaoar in the 2030s before continuing to the intersellar medium through the heliospheric nose. Quaoar is also an attractive target due to a likely disappearing methane atmosphere and cyrovolcanism. The mission studied by Brandt and his colleagues would launch using SLS and achieve 30 km/s using a Jupiter flyby. Alternatively, for an orbiter mission, a study published in 2012 concluded that Ixion and Huya are among the most feasible targets. For instance, the authors calculated that an orbiter mission could reach Ixion after 17 years cruise time if launched in 2039.

Extrasolar Kuiper belts

By 2006, astronomers had resolved dust discs thought to be Kuiper belt-like structures around nine stars other than the Sun. They appear to fall into two categories: wide belts, with radii of over 50 AU, and narrow belts (tentatively

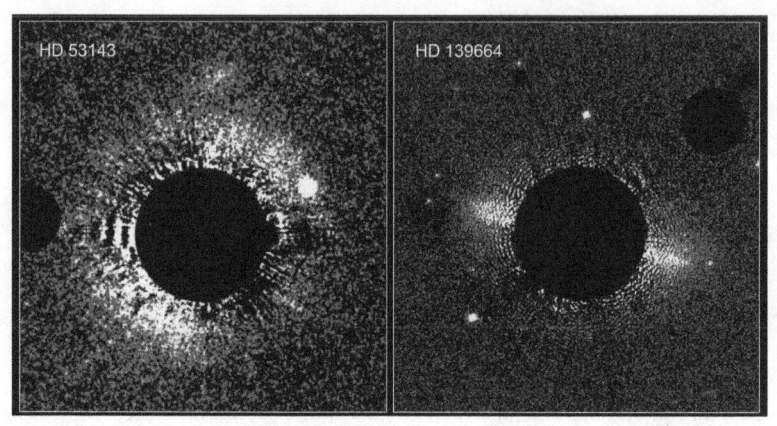

Figure 46: *Debris discs around the stars HD 139664 and HD 53143 – black circle from camera hides star to display discs.*

like that of the Solar System) with radii of between 20 and 30 AU and relatively sharp boundaries. Beyond this, 15–20% of solar-type stars have an observed infrared excess that is suggestive of massive Kuiper-belt-like structures. Most known debris discs around other stars are fairly young, but the two images on the right, taken by the *Hubble Space Telescope* in January 2006, are old enough (roughly 300 million years) to have settled into stable configurations. The left image is a "top view" of a wide belt, and the right image is an "edge view" of a narrow belt. Computer simulations of dust in the Kuiper belt suggest that when it was younger, it may have resembled the narrow rings seen around younger stars.

References

Bibliography

- Randall, Lisa (2015). *Dark Matter and the Dinosaurs*. New York: Ecco/HarperCollins Publishers. ISBN 978-0-06-232847-2.

External links

> Wikimedia Commons has media related to ***Kuiper belt***.

- Dave Jewitt's page @ UCLA[94]
  - The belt's name[95]
- List of short period comets by family[96]
- Kuiper Belt Profile[97] by NASA's Solar System Exploration[98]
- The Kuiper Belt Electronic Newsletter[99]
- Wm. Robert Johnston's TNO page[100]
- Minor Planet Center: Plot of the Outer Solar System[101], illustrating Kuiper gap
- Website of the International Astronomical Union[102] (debating the status of TNOs)
- XXVIth General Assembly 2006[103]
- nature.com article: diagram displaying inner solar system, Kuiper Belt, and Oort Cloud[104], taken from Alan Stern, S. (2003). "The evolution of comets in the Oort cloud and Kuiper belt". *Nature*. **424** (6949): 639–42. doi: 10.1038/nature01725[105]. PMID 12904784[106].
- SPACE.com: Discovery Hints at a Quadrillion Space Rocks Beyond Neptune[107] (Sara Goudarzi) August 15, 2006 06:13 am ET
- The Outer Solar System[108] Astronomy Cast episode No. 64, includes full transcript.
- The Kuiper belt[109] at 365daysofastronomy.org
- Nine Planets' webpage on the Edgeworth-Kuiper Belt and Oort Cloud[110]
- List of TNOS[111]

<indicator name="featured-star"> ⭐ </indicator>

Neptune trojan

Types of distant minor planets

- Cis-Neptunian objects
  - Centaurs
  - Neptune trojans
- Trans-Neptunian objects (TNOs)‡
  - Kuiper belt objects (KBOs)
    - Classical KBOs (cubewanos)
    - Resonant KBOs
      - Plutinos (2:3 resonance)
  - Scattered disc objects (SDOs)
    - Resonant SDOs
  - Detached objects
  - Sednoids
  - Oort cloud objects (ICO/OCOs)

‡ Trans-Neptunian dwarf planets are called "plutoids"

- $\underline{v}$
- $\underline{t}$
- $\underline{e}$[112]

Neptune trojans are bodies that orbit the Sun near one of the stable Lagrangian points of Neptune, similar to the trojans of other planets. They therefore have approximately the same orbital period as Neptune and follow roughly the same orbital path. Seventeen Neptune trojans are currently known, of which thirteen orbit near the Sun–Neptune L_4 Lagrangian point 60° ahead of Neptune and four orbit near Neptune's L_5 region 60° behind Neptune. The Neptune trojans are termed 'trojans' by analogy with the Jupiter trojans.

The discovery of 2005 TN$_{53}$ in a high-inclination (>25°) orbit was significant, because it suggested a "thick" cloud of trojans (Jupiter trojans have inclinations up to 40°), which is indicative of freeze-in capture instead of in situ or collisional formation. It is suspected that large (radius ≈ 100 km) Neptune trojans could outnumber Jupiter trojans by an order of magnitude.[113]

In 2010, the discovery of the first known L_5 Neptune trojan, 2008 LC$_{18}$, was announced. Neptune's trailing L_5 region is currently very difficult to observe because it is along the line-of-sight to the center of the Milky Way, an area of the sky crowded with stars.

It would have been possible for the *New Horizons* spacecraft to investigate 2011 HM$_{102}$, the only L_5 Neptune trojan discovered by 2014 detectable by *New Horizons*, when it passed through this region of space en route to Pluto. However, *New Horizons* may not have had sufficient downlink bandwidth, so it was decided to give precedence to the preparations for the Pluto flyby.

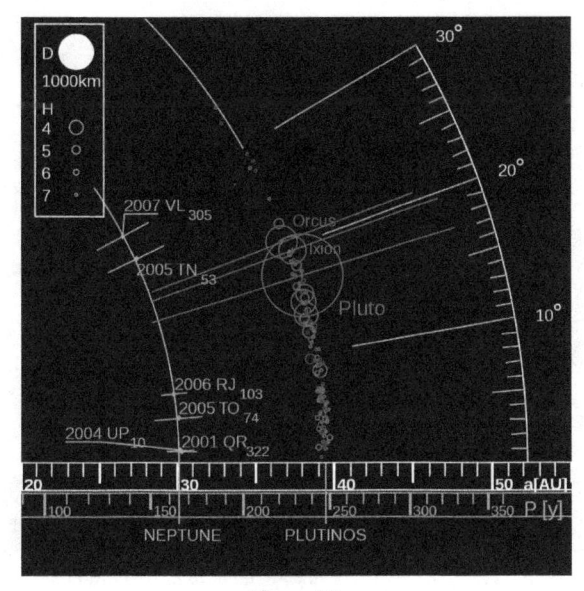

Figure 47:
Neptune's L$_4$ trojans with plutinos for reference.

Discovery and exploration

In 2001, the first Neptune trojan was discovered, 2001 QR322, near Neptune's L$_4$ region, and with it the fifth[114] known populated stable reservoir of small bodies in the Solar System. In 2005, the discovery of the high-inclination trojan 2005 TN53 has indicated that the Neptune trojans populate thick clouds, which has constrained their possible origins (see below).

On August 12, 2010, the first L$_5$ trojan, 2008 LC18, was announced. It was discovered by a dedicated survey that scanned regions where the light from the stars near the Galactic Center is obscured by dust clouds. This suggests that large L$_5$ trojans are as common as large L$_4$ trojans, to within uncertainty, further constraining models about their origins (see below).

It would have been possible for the *New Horizons* spacecraft to investigate L$_5$ Neptune trojans discovered by 2014, when it passed through this region of space en route to Pluto. Some of the patches where the light from the Galactic Center is obscured by dust clouds are along *New Horizons*'s flight path, allowing detection of objects that the spacecraft could image. 2011 HM102, the highest-inclination Neptune trojan known, was just bright enough for *New Horizons* to observe it in end-2013 at a distance of 1.2 AU. However, *New*

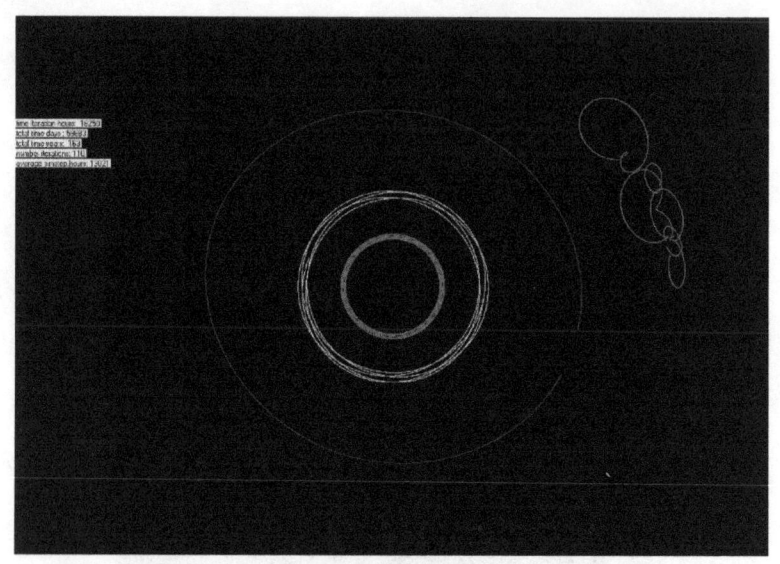

Figure 48: *An animation showing the path of six of Neptune's L_4 trojans in a rotating frame with a period equal to Neptune's orbital period. Neptune is held stationary. (Click to view.)*

Horizons may not have had sufficient downlink bandwidth, so it was eventually decided to give precedence to the preparations for the Pluto flyby.

Dynamics and origin

The orbits of Neptune trojans are highly stable; Neptune may have retained up to 50% of the original post-migration trojan population over the age of the Solar System. Neptune's L_5 can host stable trojans equally well as its L_4. Neptune trojans can librate up to 30° from their associated Lagrangian points with a 10,000-year period. Neptune trojans that escape enter orbits similar to centaurs. Although Neptune cannot currently capture stable trojans, roughly 2.8% of the centaurs within 34 AU are predicted to be Neptune co-orbitals. Of these, 54% would be in horseshoe orbits, 10% would be quasi-satellites, and 36% would be trojans (evenly split between the L_4 and L_5 groups).

The unexpected high-inclination trojans are the key to understanding the origin and evolution of the population as a whole.[115] The existence of high-inclination Neptune trojans points to a capture during planetary migration instead of in situ or collisional formation. The estimated equal number of large L_5 and L_4 trojans indicates that there was no gas drag during capture and points to a common capture mechanism for both L_4 and L_5 trojans. The capture of Neptune trojans during a migration of the planets occurs via process similar to the chaotic capture of Jupiter trojans in the Nice model. When Uranus and Neptune are near but not in a mean-motion resonance the locations where Uranus passes Neptune can circulate with a period that is in resonance with the libration periods of Neptune trojans. This results in repeated perturbations that increase the libration of existing trojans causing their orbits to become unstable. This process is reversible allowing new trojans to be captured when the planetary migration continues. For high-inclination trojans to be captured the migration must have been slow, or their inclinations must have been acquired previously.

Colors

The first four discovered Neptune trojans have similar colors. They are modestly red, slightly redder than the gray Kuiper belt objects, but not as extremely red as the high-perihelion cold classical Kuiper belt objects. This is similar to the colors of the blue lobe of the centaur color distribution, the Jupiter trojans, the irregular satellites of the gas giants, and possibly the comets, which is consistent with a similar origin of these populations of small Solar System bodies.

The Neptune trojans are too faint to efficiently observe spectroscopically with current technology, which means that a large variety of surface compositions are compatible with the observed colors.

Naming

In 2015, the IAU has adopted a new naming scheme for Neptune trojans, which are to be named after Amazons, with no differentiation between objects in L4 and L5. The Amazons were an all-female warrior tribe that fought in the Trojan War on the side of the Trojans against the Greeks. As of 2018, the only named Neptune trojan is 385571 Otrera, after Otrera, the first Amazonian queen in Greek mythology.

Members

The amount of high-inclination objects in such a small sample, in which relatively fewer high-inclination Neptune trojans are known due to observational biases, implies that high-inclination trojans may significantly outnumber low-inclination trojans. The ratio of high- to low-inclination Neptune trojans is estimated to be about 4:1. Assuming albedos of 0.05, there are an expected

400+250
−200 Neptune trojans with radii above 40 km in Neptune's L$_4$. This would indicate that large Neptune trojans are 5 to 20 times more abundant than Jupiter trojans, depending on their albedos. There may be relatively fewer smaller Neptune trojans, which could be because these fragment more readily. Large L$_5$ trojans are estimated to be as common as large L$_4$ trojans.

2001 QR$_{322}$ and 2008 LC$_{18}$ display significant dynamical instability. This means they could have been captured after planetary migration, but may as well be a long-term member that happens not to be perfectly dynamically stable.

As of March 2018, seventeen Neptune trojans are known, of which nine orbit near the Sun–Neptune L$_4$ Lagrangian point 60° ahead of Neptune, three orbit near Neptune's L$_5$ region 60° behind Neptune, and one orbits on the opposite side of Neptune (L$_3$) but frequently changes location relative to Neptune to L4 and L5. These are listed in the following table. It is constructed from the list of Neptune trojans maintained by the IAU Minor Planet Center and with diameters from Sheppard and Trujillo's paper on 2008 LC$_{18}$, unless otherwise noted.

Name	Prov. designation	Lagrangian point	q (AU)	Q (AU)	i (°)	Abs. mag	Diameter (km)	Year of identification	Notes	MPC
2001 QR$_{322}$		L$_4$	29.404	31.011	1.3	8.2	~140	2001	First Neptune trojan discovered	MPC[116]
385571 Otrera	2004 UP$_{10}$	L$_4$	29.318	30.942	1.4	8.8	~100	2004	First Neptune trojan numbered and named	MPC[117]
2005 TN$_{53}$		L$_4$	28.092	32.162	25.0	9.0	~80	2005	First high-inclination trojan discovered	MPC[118]
(385695) 2005 TO$_{74}$		L$_4$	28.469	31.771	5.3	8.5	~100	2005	–	MPC[119]
2006 RJ$_{103}$		L$_4$	29.077	31.014	8.2	7.5	~180	2006	–	MPC[120]
2007 VL$_{305}$		L$_4$	28.130	32.028	28.1	8.0	~160	2007	–	MPC[121]
2008 LC$_{18}$		L$_5$	27.365	32.479	27.6	8.4	~100	2008	First L$_5$ trojan discovered	MPC[122]
2004 KV$_{18}$		L$_5$	24.553	35.851	13.6	8.9	56	2011	Temporary Neptune trojan	MPC[123]

(316179) 2010 EN$_{65}$	L$_3$	21.109	40.613	19.2	6.9	~200	–	Jumping trojan	MPC[124]
2010 TS$_{191}$	L$_4$	28.608	31.253	6.6	8.1	~120	2016	Announced on 2016/05/31	MPC[125]
2010 TT$_{191}$	L$_4$	27.913	32.189	4.3	8.0	~130	2016	Announced on 2016/05/31	MPC[126]
2011 HM$_{102}$	L$_5$	27.662	32.455	29.4	8.1	90–180	2012	–	MPC[127]
2011 SO$_{277}$	L$_4$	29.622	30.503	9.6	7.7	~140	2016	Announced on 2016/05/31	MPC[128]
2011 WG$_{157}$	L$_4$	29.064	30.878	22.3	7.1	~170	2016	Announced on 2016/05/31	MPC[129]
2012 UV$_{177}$	L$_4$	27.806	32.259	20.8	9.2	~80	–	–	MPC[130]
2013 KY$_{18}$	L$_5$	26.598	33.873	6.7	6.8	~200	2016	Announced on 2016/-05/31, stability uncertain	MPC[131]
2014 QO$_{441}$	L$_4$	26.961	33.215	18.8	8.2	~130	–	Most eccentric stable Neptune trojan	MPC[132]
2014 QP$_{441}$	L$_4$	28.022	32.110	19.4	9.1	~90	–	–	MPC[133]

2005 TN$_{74}$[134] and (309239) 2007 RW$_{10}$ were thought to be Neptune trojans at the time of their discovery, but further observations have disconfirmed their membership. 2005 TN$_{74}$ is currently thought to be in a 3:5 resonance with Neptune. (309239) 2007 RW$_{10}$ is currently following a quasi-satellite loop around Neptune.

External links

- Horner, J.; Lykawka, P. S. (2010). "Planetary Trojans – the main source of short period comets?". *International Journal of Astrobiology*. **9** (4): 227–234. arXiv:1007.2541[135] ⓐ. Bibcode: 2010IJAsB...9..227H[136]. doi: 10.1017/S1473550410000212[137].
- "New Trojan asteroid hints at huge Neptunian cloud"[138]. New Scientist.

Formation and migration

Formation and evolution of the Solar System

The formation and evolution of the Solar System began 4.6 billion years ago with the gravitational collapse of a small part of a giant molecular cloud. Most of the collapsing mass collected in the center, forming the Sun, while the rest flattened into a protoplanetary disk out of which the planets, moons, asteroids, and other small Solar System bodies formed.

This model, known as the nebular hypothesis and now refined as the Nice model (2005), was first developed in the 18th century by Emanuel Swedenborg, Immanuel Kant, and Pierre-Simon Laplace. Its subsequent development has interwoven a variety of scientific disciplines including astronomy, physics, geology, and planetary science. Since the dawn of the space age in the 1950s and the discovery of extrasolar planets in the 1990's, the model has been both challenged and refined to account for new observations.

The Solar System has evolved considerably since its initial formation. Many moons have formed from circling discs of gas and dust around their parent planets, while other moons are thought to have formed independently and later been captured by their planets. Still others, such as Earth's Moon, may be the result of giant collisions. Collisions between bodies have occurred continually up to the present day and have been central to the evolution of the Solar System. The positions of the planets often shifted due to gravitational interactions. This planetary migration is now thought to have been responsible for much of the Solar System's early evolution.

In roughly 5 billion years, the Sun will cool and expand outward to many times its current diameter (becoming a red giant), before casting off its outer layers as a planetary nebula and leaving behind a stellar remnant known as a white dwarf. In the far distant future, the gravity of passing stars will gradually reduce the

Figure 49: *Artist's conception of a protoplanetary disk*

Sun's retinue of planets. Some planets will be destroyed, others ejected into interstellar space. Ultimately, over the course of tens of billions of years, it is likely that the Sun will be left with none of the original bodies in orbit around it.

History

Ideas concerning the origin and fate of the world date from the earliest known writings; however, for almost all of that time, there was no attempt to link such theories to the existence of a "Solar System", simply because it was not generally thought that the Solar System, in the sense we now understand it, existed. The first step toward a theory of Solar System formation and evolution was the general acceptance of heliocentrism, which placed the Sun at the centre of the system and the Earth in orbit around it. This concept had developed for millennia (Aristarchus of Samos had suggested it as early as 250 BC), but was not widely accepted until the end of the 17th century. The first recorded use of the term "Solar System" dates from 1704.

The current standard theory for Solar System formation, the nebular hypothesis, has fallen into and out of favour since its formulation by Emanuel Swedenborg, Immanuel Kant, and Pierre-Simon Laplace in the 18th century. The most significant criticism of the hypothesis was its apparent inability to explain the Sun's relative lack of angular momentum when compared to the planets. However, since the early 1980s studies of young stars have shown them to be surrounded by cool discs of dust and gas, exactly as the nebular hypothesis predicts, which has led to its re-acceptance.

Figure 50: *Pierre-Simon Laplace, one of the originators of the nebular hypothesis*

Understanding of how the Sun is expected to continue to evolve required an understanding of the source of its power. Arthur Stanley Eddington's confirmation of Albert Einstein's theory of relativity led to his realisation that the Sun's energy comes from nuclear fusion reactions in its core, fusing hydrogen into helium. In 1935, Eddington went further and suggested that other elements also might form within stars. Fred Hoyle elaborated on this premise by arguing that evolved stars called red giants created many elements heavier than hydrogen and helium in their cores. When a red giant finally casts off its outer layers, these elements would then be recycled to form other star systems.

Formation

Pre-solar nebula

The nebular hypothesis says that the Solar System formed from the gravitational collapse of a fragment of a giant molecular cloud. The cloud was about 20 parsec (65 light years) across, while the fragments were roughly 1 parsec (three and a quarter light-years) across. The further collapse of the fragments led to the formation of dense cores 0.01–0.1 pc (2,000–20,000 AU) in size.[139] One of these collapsing fragments (known as the *pre-solar nebula*) formed what became the Solar System. The composition of this region with a mass

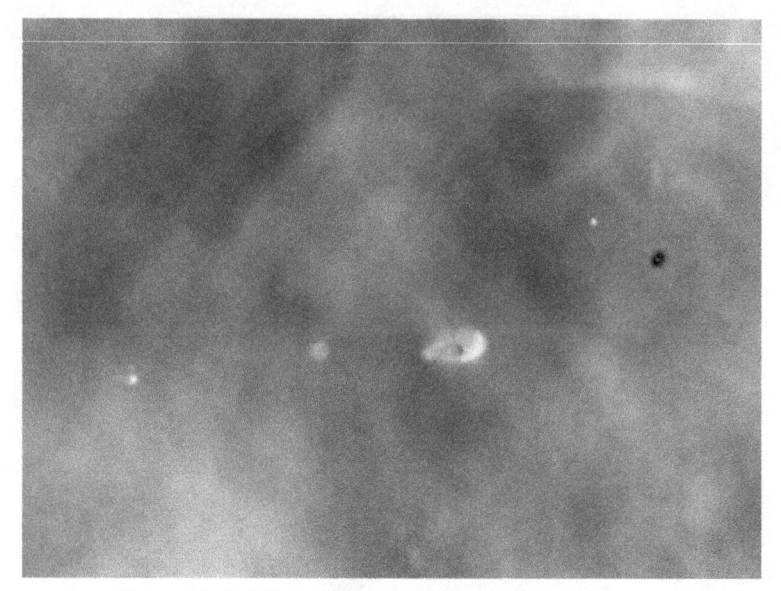

Figure 51: *Hubble image of protoplanetary discs in the Orion Nebula, a light-years-wide "stellar nursery" probably very similar to the primordial nebula from which the Sun formed*

just over that of the Sun ($M_\odot$) was about the same as that of the Sun today, with hydrogen, along with helium and trace amounts of lithium produced by Big Bang nucleosynthesis, forming about 98% of its mass. The remaining 2% of the mass consisted of heavier elements that were created by nucleosynthesis in earlier generations of stars.[140] Late in the life of these stars, they ejected heavier elements into the interstellar medium.

The oldest inclusions found in meteorites, thought to trace the first solid material to form in the pre-solar nebula, are 4568.2 million years old, which is one definition of the age of the Solar System. Studies of ancient meteorites reveal traces of stable daughter nuclei of short-lived isotopes, such as iron-60, that only form in exploding, short-lived stars. This indicates that one or more supernovae occurred near the Sun while it was forming. A shock wave from a supernova may have triggered the formation of the Sun by creating relatively dense regions within the cloud, causing these regions to collapse. Because only massive, short-lived stars produce supernovae, the Sun must have formed in a large star-forming region that produced massive stars, possibly similar to the Orion Nebula. Studies of the structure of the Kuiper belt and of anomalous materials within it suggest that the Sun formed within a cluster of between 1,000 and 10,000 stars with a diameter of between 6.5 and 19.5 light years and a collective mass of 3,000 $M_\odot$. This cluster began to break apart between 135 million and 535 million years after formation. Several simulations of our young Sun interacting with close-passing stars over the first 100 million years of its life produce anomalous orbits observed in the outer Solar System, such as detached objects.

Because of the conservation of angular momentum, the nebula spun faster as it collapsed. As the material within the nebula condensed, the atoms within it began to collide with increasing frequency, converting their kinetic energy into heat. The centre, where most of the mass collected, became increasingly hotter than the surrounding disc. Over about 100,000 years, the competing forces of gravity, gas pressure, magnetic fields, and rotation caused the contracting nebula to flatten into a spinning protoplanetary disc with a diameter of about 200 AU and form a hot, dense protostar (a star in which hydrogen fusion has not yet begun) at the centre.

At this point in its evolution, the Sun is thought to have been a T Tauri star. Studies of T Tauri stars show that they are often accompanied by discs of pre-planetary matter with masses of 0.001–0.1 $M_\odot$. These discs extend to several hundred AU—the Hubble Space Telescope has observed protoplanetary discs of up to 1000 AU in diameter in star-forming regions such as the Orion Nebula—and are rather cool, reaching a surface temperature of only about 1000 kelvin at their hottest. Within 50 million years, the temperature and pressure at the core of the Sun became so great that its hydrogen began to fuse, creating an internal source of energy that countered gravitational contraction until hydrostatic equilibrium was achieved. This marked the Sun's entry into the prime phase of its life, known as the main sequence. Main-sequence stars derive energy from the fusion of hydrogen into helium in their cores. The Sun remains a main-sequence star today.

Formation of the planets

The various planets are thought to have formed from the solar nebula, the disc-shaped cloud of gas and dust left over from the Sun's formation. The currently accepted method by which the planets formed is accretion, in which the planets began as dust grains in orbit around the central protostar. Through direct contact, these grains formed into clumps up to 200 metres in diameter, which in turn collided to form larger bodies (planetesimals) of ∼10 kilometres (km) in size. These gradually increased through further collisions, growing at the rate of centimetres per year over the course of the next few million years.

The inner Solar System, the region of the Solar System inside 4 AU, was too warm for volatile molecules like water and methane to condense, so the planetesimals that formed there could only form from compounds with high

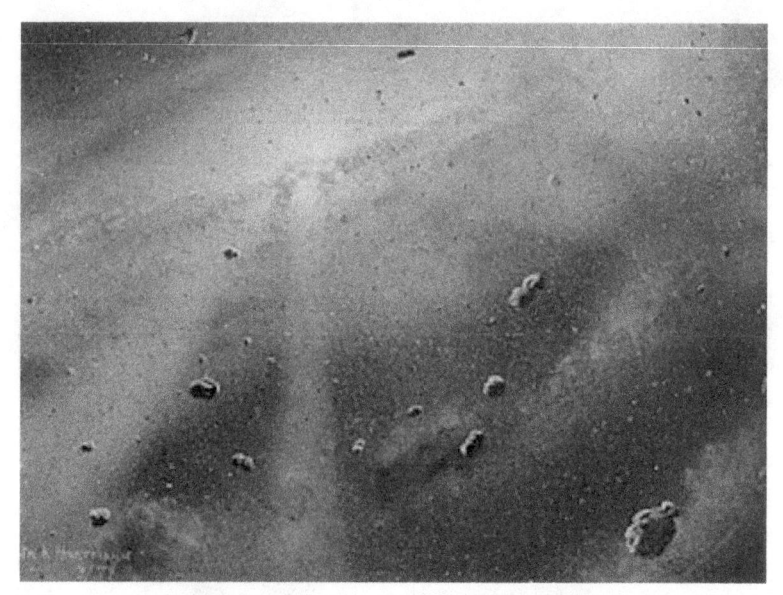

Figure 52: *Artist's conception of the solar nebula*

melting points, such as metals (like iron, nickel, and aluminium) and rocky silicates. These rocky bodies would become the terrestrial planets (Mercury, Venus, Earth, and Mars). These compounds are quite rare in the Universe, comprising only 0.6% of the mass of the nebula, so the terrestrial planets could not grow very large. The terrestrial embryos grew to about 0.05 Earth masses ($M_\oplus$) and ceased accumulating matter about 100,000 years after the formation of the Sun; subsequent collisions and mergers between these planet-sized bodies allowed terrestrial planets to grow to their present sizes (see Terrestrial planets below).

When the terrestrial planets were forming, they remained immersed in a disk of gas and dust. The gas was partially supported by pressure and so did not orbit the Sun as rapidly as the planets. The resulting drag and, more importantly, gravitational interactions with the surrounding material caused a transfer of angular momentum, and as a result the planets gradually migrated to new orbits. Models show that density and temperature variations in the disk governed this rate of migration, but the net trend was for the inner planets to migrate inward as the disk dissipated, leaving the planets in their current orbits.

The giant planets (Jupiter, Saturn, Uranus, and Neptune) formed further out, beyond the frost line, which is the point between the orbits of Mars and Jupiter where the material is cool enough for volatile icy compounds to remain solid.

The ices that formed the Jovian planets were more abundant than the metals and silicates that formed the terrestrial planets, allowing the giant planets to grow massive enough to capture hydrogen and helium, the lightest and most abundant elements. Planetesimals beyond the frost line accumulated up to 4 $M_\oplus$ within about 3 million years. Today, the four giant planets comprise just under 99% of all the mass orbiting the Sun.[141] Theorists believe it is no accident that Jupiter lies just beyond the frost line. Because the frost line accumulated large amounts of water via evaporation from infalling icy material, it created a region of lower pressure that increased the speed of orbiting dust particles and halted their motion toward the Sun. In effect, the frost line acted as a barrier that caused material to accumulate rapidly at $\sim$5 AU from the Sun. This excess material coalesced into a large embryo (or core) on the order of 10 $M_\oplus$, which began to accumulate an envelope via accretion of gas from the surrounding disc at an ever-increasing rate. Once the envelope mass became about equal to the solid core mass, growth proceeded very rapidly, reaching about 150 Earth masses $\sim 10^5$ years thereafter and finally topping out at 318 $M_\oplus$. Saturn may owe its substantially lower mass simply to having formed a few million years after Jupiter, when there was less gas available to consume.

T Tauri stars like the young Sun have far stronger stellar winds than more stable, older stars. Uranus and Neptune are thought to have formed after Jupiter and Saturn did, when the strong solar wind had blown away much of the disc material. As a result, the planets accumulated little hydrogen and helium—not more than 1 $M_\oplus$ each. Uranus and Neptune are sometimes referred to as failed cores. The main problem with formation theories for these planets is the timescale of their formation. At the current locations it would have taken millions of years for their cores to accrete. This means that Uranus and Neptune may have formed closer to the Sun—near or even between Jupiter and Saturn—and later migrated or were ejected outward (see Planetary migration below). Motion in the planetesimal era was not all inward toward the Sun; the *Stardust* sample return from Comet Wild 2 has suggested that materials from the early formation of the Solar System migrated from the warmer inner Solar System to the region of the Kuiper belt.

After between three and ten million years, the young Sun's solar wind would have cleared away all the gas and dust in the protoplanetary disc, blowing it into interstellar space, thus ending the growth of the planets.

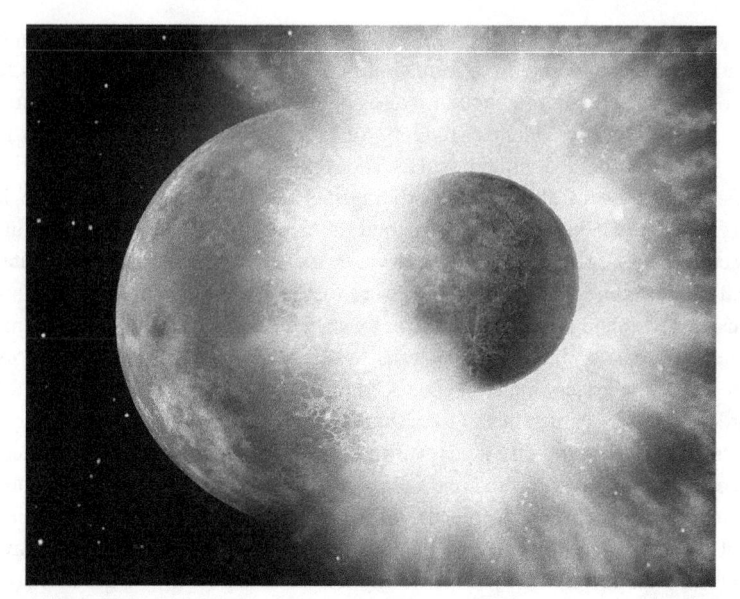

Figure 53: *Artist's conception of the giant impact thought to have formed the Moon*

Subsequent evolution

The planets were originally thought to have formed in or near their current orbits. From that a minimum mass of the nebula i.e. the protoplanetary disc, was derived that was necessary to form the planets - the minimum mass solar nebula. It was derived that the nebula mass must have exceeded 3585 times that of the Earth.

However, this has been questioned during the last 20 years. Currently, many planetary scientists think that the Solar System might have looked very different after its initial formation: several objects at least as massive as Mercury were present in the inner Solar System, the outer Solar System was much more compact than it is now, and the Kuiper belt was much closer to the Sun.

Terrestrial planets

At the end of the planetary formation epoch the inner Solar System was populated by 50–100 Moon- to Mars-sized planetary embryos. Further growth was possible only because these bodies collided and merged, which took less than 100 million years. These objects would have gravitationally interacted with one another, tugging at each other's orbits until they collided, growing larger until the four terrestrial planets we know today took shape. One such giant collision is thought to have formed the Moon (see Moons below), while another removed the outer envelope of the young Mercury.

One unresolved issue with this model is that it cannot explain how the initial orbits of the proto-terrestrial planets, which would have needed to be highly eccentric to collide, produced the remarkably stable and nearly circular orbits they have today. One hypothesis for this "eccentricity dumping" is that the terrestrials formed in a disc of gas still not expelled by the Sun. The "gravitational drag" of this residual gas would have eventually lowered the planets' energy, smoothing out their orbits. However, such gas, if it existed, would have prevented the terrestrial planets' orbits from becoming so eccentric in the first place. Another hypothesis is that gravitational drag occurred not between the planets and residual gas but between the planets and the remaining small bodies. As the large bodies moved through the crowd of smaller objects, the smaller objects, attracted by the larger planets' gravity, formed a region of higher density, a "gravitational wake", in the larger objects' path. As they did so, the increased gravity of the wake slowed the larger objects down into more regular orbits.

Asteroid belt

The outer edge of the terrestrial region, between 2 and 4 AU from the Sun, is called the asteroid belt. The asteroid belt initially contained more than enough matter to form 2–3 Earth-like planets, and, indeed, a large number of planetesimals formed there. As with the terrestrials, planetesimals in this region later coalesced and formed 20–30 Moon- to Mars-sized planetary embryos; however, the proximity of Jupiter meant that after this planet formed, 3 million years after the Sun, the region's history changed dramatically. Orbital resonances with Jupiter and Saturn are particularly strong in the asteroid belt, and gravitational interactions with more massive embryos scattered many planetesimals into those resonances. Jupiter's gravity increased the velocity of objects within these resonances, causing them to shatter upon collision with other bodies, rather than accrete.

As Jupiter migrated inward following its formation (see Planetary migration below), resonances would have swept across the asteroid belt, dynamically exciting the region's population and increasing their velocities relative to each other. The cumulative action of the resonances and the embryos either scattered the planetesimals away from the asteroid belt or excited their orbital inclinations and eccentricities. Some of those massive embryos too were ejected by Jupiter, while others may have migrated to the inner Solar System and played a role in the final accretion of the terrestrial planets. During this primary depletion period, the effects of the giant planets and planetary embryos left the

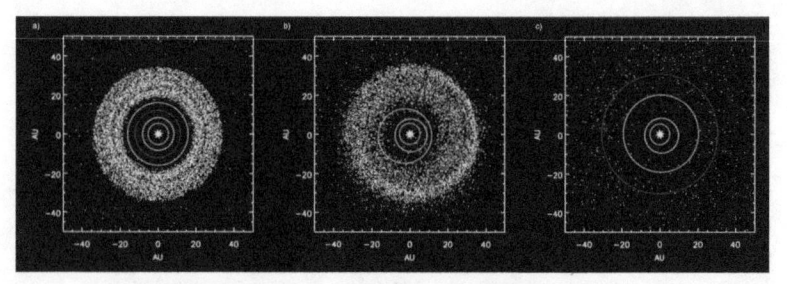

Figure 54: *Simulation showing outer planets and Kuiper belt: a) Before Jupiter/Saturn 2:1 resonance b) Scattering of Kuiper belt objects into the Solar System after the orbital shift of Neptune c) After ejection of Kuiper belt bodies by Jupiter*

asteroid belt with a total mass equivalent to less than 1% that of the Earth, composed mainly of small planetesimals. This is still 10–20 times more than the current mass in the main belt, which is now about 1/2,000 $M_\oplus$. A secondary depletion period that brought the asteroid belt down close to its present mass is thought to have followed when Jupiter and Saturn entered a temporary 2:1 orbital resonance (see below).

The inner Solar System's period of giant impacts probably played a role in the Earth acquiring its current water content ($\sim 6 \times 10^{21}$ kg) from the early asteroid belt. Water is too volatile to have been present at Earth's formation and must have been subsequently delivered from outer, colder parts of the Solar System. The water was probably delivered by planetary embryos and small planetesimals thrown out of the asteroid belt by Jupiter. A population of main-belt comets discovered in 2006 has been also suggested as a possible source for Earth's water. In contrast, comets from the Kuiper belt or farther regions delivered not more than about 6% of Earth's water. The panspermia hypothesis holds that life itself may have been deposited on Earth in this way, although this idea is not widely accepted.

Planetary migration

According to the nebular hypothesis, the outer two planets may be in the "wrong place". Uranus and Neptune (known as the "ice giants") exist in a region where the reduced density of the solar nebula and longer orbital times render their formation highly implausible. The two are instead thought to have formed in orbits near Jupiter and Saturn, where more material was available, and to have migrated outward to their current positions over hundreds of millions of years.

The migration of the outer planets is also necessary to account for the existence and properties of the Solar System's outermost regions. Beyond Neptune, the

Solar System continues into the Kuiper belt, the scattered disc, and the Oort cloud, three sparse populations of small icy bodies thought to be the points of origin for most observed comets. At their distance from the Sun, accretion was too slow to allow planets to form before the solar nebula dispersed, and thus the initial disc lacked enough mass density to consolidate into a planet. The Kuiper belt lies between 30 and 55 AU from the Sun, while the farther scattered disc extends to over 100 AU, and the distant Oort cloud begins at about 50,000 AU. Originally, however, the Kuiper belt was much denser and closer to the Sun, with an outer edge at approximately 30 AU. Its inner edge would have been just beyond the orbits of Uranus and Neptune, which were in turn far closer to the Sun when they formed (most likely in the range of 15–20 AU), and in 50% of simulations ended up opposite locations, with Uranus farther from the Sun than Neptune.

According to the Nice model, after the formation of the Solar System, the orbits of all the giant planets continued to change slowly, influenced by their interaction with the large number of remaining planetesimals. After 500–600 million years (about 4 billion years ago) Jupiter and Saturn fell into a 2:1 resonance: Saturn orbited the Sun once for every two Jupiter orbits. This resonance created a gravitational push against the outer planets, possibly causing Neptune to surge past Uranus and plough into the ancient Kuiper belt. The planets scattered the majority of the small icy bodies inwards, while themselves moving outwards. These planetesimals then scattered off the next planet they encountered in a similar manner, moving the planets' orbits outwards while they moved inwards. This process continued until the planetesimals interacted with Jupiter, whose immense gravity sent them into highly elliptical orbits or even ejected them outright from the Solar System. This caused Jupiter to move slightly inward.[142] Those objects scattered by Jupiter into highly elliptical orbits formed the Oort cloud; those objects scattered to a lesser degree by the migrating Neptune formed the current Kuiper belt and scattered disc. This scenario explains the Kuiper belt's and scattered disc's present low mass. Some of the scattered objects, including Pluto, became gravitationally tied to Neptune's orbit, forcing them into mean-motion resonances. Eventually, friction within the planetesimal disc made the orbits of Uranus and Neptune circular again.

In contrast to the outer planets, the inner planets are not thought to have migrated significantly over the age of the Solar System, because their orbits have remained stable following the period of giant impacts.

Another question is why Mars came out so small compared with Earth. A study by Southwest Research Institute, San Antonio, Texas, published June 6, 2011 (called the Grand tack hypothesis), proposes that Jupiter had migrated inward to 1.5 AU. After Saturn formed, migrated inward, and established the 2:3 mean motion resonance with Jupiter, the study assumes that both planets migrated back to their present positions. Jupiter thus would have consumed much of the material that would have created a bigger Mars. The same simulations also reproduce the characteristics of the modern asteroid belt, with dry asteroids and water-rich objects similar to comets. However, it is unclear whether conditions in the solar nebula would have allowed Jupiter and Saturn to move back to their current positions, and according to current estimates this possibility appears unlikely. Moreover, alternative explanations for the small mass of Mars exist.

Late Heavy Bombardment and after

Life timeline

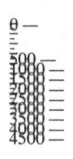

Figure 55: *Meteor Crater in Arizona. Created 50,000 years ago by an impactor about 50 metres (160 ft) across, it shows that the accretion of the Solar System is not over.*

Axis scale: million years

Also see: *Human timeline* and *Nature timeline*

Gravitational disruption from the outer planets' migration would have sent large numbers of asteroids into the inner Solar System, severely depleting the original belt until it reached today's extremely low mass. This event may have triggered the Late Heavy Bombardment that occurred approximately 4 billion years ago, 500–600 million years after the formation of the Solar System. This period of heavy bombardment lasted several hundred million years and is evident in the cratering still visible on geologically dead bodies of the inner Solar System such as the Moon and Mercury. The oldest known evidence for life on Earth dates to 3.8 billion years ago—almost immediately after the end of the Late Heavy Bombardment.

Impacts are thought to be a regular (if currently infrequent) part of the evolution of the Solar System. That they continue to happen is evidenced by the collision of Comet Shoemaker–Levy 9 with Jupiter in 1994, the 2009 Jupiter impact event, the Tunguska event, the Chelyabinsk meteor and the impact feature Meteor Crater in Arizona. The process of accretion, therefore, is not complete, and may still pose a threat to life on Earth.

Over the course of the Solar System's evolution, comets were ejected out of the inner Solar System by the gravity of the giant planets, and sent thousands of AU outward to form the Oort cloud, a spherical outer swarm of cometary nuclei at the farthest extent of the Sun's gravitational pull. Eventually, after about 800 million years, the gravitational disruption caused by galactic tides, passing stars and giant molecular clouds began to deplete the cloud, sending comets into the inner Solar System. The evolution of the outer Solar System also appears to have been influenced by space weathering from the solar wind, micrometeorites, and the neutral components of the interstellar medium.

The evolution of the asteroid belt after Late Heavy Bombardment was mainly governed by collisions. Objects with large mass have enough gravity to retain any material ejected by a violent collision. In the asteroid belt this usually is not the case. As a result, many larger objects have been broken apart, and sometimes newer objects have been forged from the remnants in less violent collisions. Moons around some asteroids currently can only be explained as consolidations of material flung away from the parent object without enough energy to entirely escape its gravity.

Moons

Moons have come to exist around most planets and many other Solar System bodies. These natural satellites originated by one of three possible mechanisms:

- Co-formation from a circumplanetary disc (only in the cases of the giant planets);
- Formation from impact debris (given a large enough impact at a shallow angle); and
- Capture of a passing object.

Jupiter and Saturn have several large moons, such as Io, Europa, Ganymede and Titan, which may have originated from discs around each giant planet in much the same way that the planets formed from the disc around the Sun.[143] This origin is indicated by the large sizes of the moons and their proximity to the planet. These attributes are impossible to achieve via capture, while the gaseous nature of the primaries also make formation from collision debris unlikely. The outer moons of the giant planets tend to be small and have eccentric orbits with arbitrary inclinations. These are the characteristics expected of captured bodies. Most such moons orbit in the direction opposite the rotation of their primary. The largest irregular moon is Neptune's moon Triton, which is thought to be a captured Kuiper belt object.

Moons of solid Solar System bodies have been created by both collisions and capture. Mars's two small moons, Deimos and Phobos, are thought to be captured asteroids.[144] The Earth's Moon is thought to have formed as a result of a single, large head-on collision. The impacting object probably had a mass comparable to that of Mars, and the impact probably occurred near the end of the period of giant impacts. The collision kicked into orbit some of the impactor's mantle, which then coalesced into the Moon. The impact was probably the last in the series of mergers that formed the Earth. It has been further hypothesized that the Mars-sized object may have formed at one of the stable Earth–Sun Lagrangian points (either L_4 or L_5) and drifted from its position. The moons of trans-Neptunian objects Pluto (Charon) and Orcus (Vanth) may also have formed by means of a large collision: the Pluto–Charon, Orcus–Vanth and Earth–Moon systems are unusual in the Solar System in that the satellite's mass is at least 1% that of the larger body.

Future

Astronomers estimate that the Solar System as we know it today will not change drastically until the Sun has fused almost all the hydrogen fuel in its core into helium, beginning its evolution from the main sequence of the Hertzsprung–Russell diagram and into its red-giant phase. Even so, the Solar System will continue to evolve until then.

Long-term stability

The Solar System is chaotic over million- and billion-year timescales, with the orbits of the planets open to long-term variations. One notable example of this chaos is the Neptune–Pluto system, which lies in a 3:2 orbital resonance. Although the resonance itself will remain stable, it becomes impossible to predict the position of Pluto with any degree of accuracy more than 10–20 million years (the Lyapunov time) into the future. Another example is Earth's axial tilt, which, due to friction raised within Earth's mantle by tidal interactions with the Moon (see below), will be incomputable at some point between 1.5 and 4.5 billion years from now.

The outer planets' orbits are chaotic over longer timescales, with a Lyapunov time in the range of 2–230 million years. In all cases this means that the position of a planet along its orbit ultimately becomes impossible to predict with any certainty (so, for example, the timing of winter and summer become uncertain), but in some cases the orbits themselves may change dramatically. Such chaos manifests most strongly as changes in eccentricity, with some planets' orbits becoming significantly more—or less—elliptical.

Ultimately, the Solar System is stable in that none of the planets are likely to collide with each other or be ejected from the system in the next few billion years. Beyond this, within five billion years or so Mars's eccentricity may grow to around 0.2, such that it lies on an Earth-crossing orbit, leading to a potential collision. In the same timescale, Mercury's eccentricity may grow even further, and a close encounter with Venus could theoretically eject it from the Solar System altogether or send it on a collision course with Venus or Earth. This could happen within a billion years, according to numerical simulations in which Mercury's orbit is perturbed.

Moon–ring systems

The evolution of moon systems is driven by tidal forces. A moon will raise a tidal bulge in the object it orbits (the primary) due to the differential gravitational force across diameter of the primary. If a moon is revolving in the same direction as the planet's rotation and the planet is rotating faster than the orbital period of the moon, the bulge will constantly be pulled ahead of the moon. In this situation, angular momentum is transferred from the rotation of the primary to the revolution of the satellite. The moon gains energy and gradually spirals outward, while the primary rotates more slowly over time.

The Earth and its Moon are one example of this configuration. Today, the Moon is tidally locked to the Earth; one of its revolutions around the Earth (currently about 29 days) is equal to one of its rotations about its axis, so it always shows one face to the Earth. The Moon will continue to recede from Earth, and Earth's spin will continue to slow gradually. In about 50 billion years, if they survive the Sun's expansion, the Earth and Moon will become tidally locked to each other; each will be caught up in what is called a "spin–orbit resonance" in which the Moon will circle the Earth in about 47 days and both Moon and Earth will rotate around their axes in the same time, each only visible from one hemisphere of the other. Other examples are the Galilean moons of Jupiter (as well as many of Jupiter's smaller moons) and most of the larger moons of Saturn.

A different scenario occurs when the moon is either revolving around the primary faster than the primary rotates, or is revolving in the direction opposite the planet's rotation. In these cases, the tidal bulge lags behind the moon in its orbit. In the former case, the direction of angular momentum transfer is reversed, so the rotation of the primary speeds up while the satellite's orbit shrinks. In the latter case, the angular momentum of the rotation and revolution have opposite signs, so transfer leads to decreases in the magnitude of each (that cancel each other out).[145] In both cases, tidal deceleration causes the moon to spiral in towards the primary until it either is torn apart by tidal stresses, potentially creating a planetary ring system, or crashes into the planet's surface or atmosphere. Such a fate awaits the moons Phobos of Mars (within 30 to 50 million years), Triton of Neptune (in 3.6 billion years), Metis and

Formation and evolution of the Solar System

Figure 56: *Neptune and its moon Triton, taken by Voyager 2. Triton's orbit will eventually take it within Neptune's Roche limit, tearing it apart and possibly forming a new ring system.*

Adrastea of Jupiter, and at least 16 small satellites of Uranus and Neptune. Uranus's Desdemona may even collide with one of its neighboring moons.[146]

A third possibility is where the primary and moon are tidally locked to each other. In that case, the tidal bulge stays directly under the moon, there is no transfer of angular momentum, and the orbital period will not change. Pluto and Charon are an example of this type of configuration.

Prior to the 2004 arrival of the *Cassini–Huygens* spacecraft, the rings of Saturn were widely thought to be much younger than the Solar System and were not expected to survive beyond another 300 million years. Gravitational interactions with Saturn's moons were expected to gradually sweep the rings' outer edge toward the planet, with abrasion by meteorites and Saturn's gravity eventually taking the rest, leaving Saturn unadorned. However, data from the *Cassini* mission led scientists to revise that early view. Observations revealed 10 km-wide icy clumps of material that repeatedly break apart and reform, keeping the rings fresh. Saturn's rings are far more massive than the rings of the other giant planets. This large mass is thought to have preserved Saturn's rings since it first formed 4.5 billion years ago, and is likely to preserve them for billions of years to come.

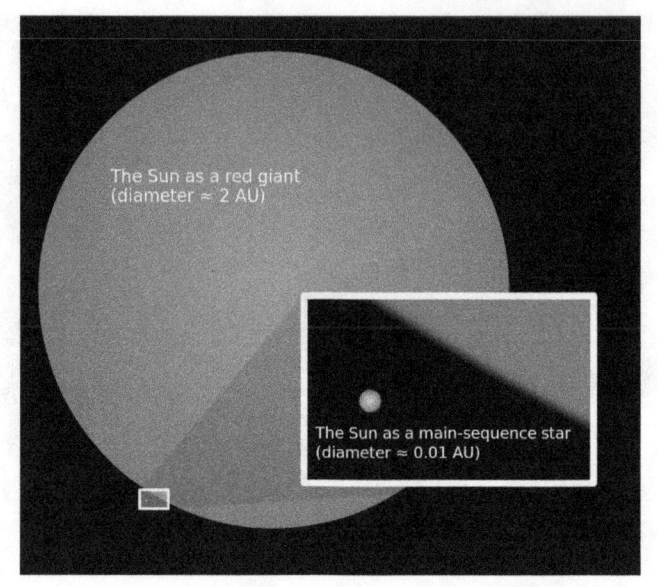

Figure 57: *Relative size of the Sun as it is now (inset) compared to its estimated future size as a red giant*

The Sun and planetary environments

In the long term, the greatest changes in the Solar System will come from changes in the Sun itself as it ages. As the Sun burns through its supply of hydrogen fuel, it gets hotter and burns the remaining fuel even faster. As a result, the Sun is growing brighter at a rate of ten percent every 1.1 billion years. In one billion years' time, as the Sun's radiation output increases, its circumstellar habitable zone will move outwards, making the Earth's surface too hot for liquid water to exist there naturally. At this point, all life on land will become extinct. Evaporation of water, a potent greenhouse gas, from the oceans' surface could accelerate temperature increase, potentially ending all life on Earth even sooner. During this time, it is possible that as Mars's surface temperature gradually rises, carbon dioxide and water currently frozen under the surface regolith will release into the atmosphere, creating a greenhouse effect that will heat the planet until it achieves conditions parallel to Earth today, providing a potential future abode for life. By 3.5 billion years from now, Earth's surface conditions will be similar to those of Venus today.

Around 5.4 billion years from now, the core of the Sun will become hot enough to trigger hydrogen fusion in its surrounding shell. This will cause the outer layers of the star to expand greatly, and the star will enter a phase of its life in which it is called a red giant.[147] Within 7.5 billion years, the Sun will have expanded to a radius of 1.2 AU—256 times its current size. At the tip of the red giant branch, as a result of the vastly increased surface area, the Sun's surface will be much cooler (about 2600 K) than now and its luminosity much higher—up to 2,700 current solar luminosities. For part of its red giant life, the Sun will have a strong stellar wind that will carry away around 33% of its mass.[148] During these times, it is possible that Saturn's moon Titan could achieve surface temperatures necessary to support life.

As the Sun expands, it will swallow the planets Mercury and Venus. Earth's fate is less clear; although the Sun will envelop Earth's current orbit, the star's loss of mass (and thus weaker gravity) will cause the planets' orbits to move farther out. If it were only for this, Venus and Earth would probably escape incineration, but a 2008 study suggests that Earth will likely be swallowed up as a result of tidal interactions with the Sun's weakly bound outer envelope.

Gradually, the hydrogen burning in the shell around the solar core will increase the mass of the core until it reaches about 45% of the present solar mass. At this point the density and temperature will become so high that the fusion of helium into carbon will begin, leading to a helium flash; the Sun will shrink from around 250 to 11 times its present (main-sequence) radius. Consequently, its luminosity will decrease from around 3,000 to 54 times its current level, and its surface temperature will increase to about 4770 K. The Sun will become a horizontal giant, burning helium in its core in a stable fashion much like it burns hydrogen today. The helium-fusing stage will last only 100 million years. Eventually, it will have to again resort to the reserves of hydrogen and helium in its outer layers and will expand a second time, turning into what is known as an asymptotic giant. Here the luminosity of the Sun will increase again, reaching about 2,090 present luminosities, and it will cool to about 3500 K. This phase lasts about 30 million years, after which, over the course of a further 100,000 years, the Sun's remaining outer layers will fall away, ejecting a vast stream of matter into space and forming a halo known (misleadingly) as a planetary nebula. The ejected material will contain the helium and carbon produced by the Sun's nuclear reactions, continuing the enrichment of the interstellar medium with heavy elements for future generations of stars.

This is a relatively peaceful event, nothing akin to a supernova, which the Sun is too small to undergo as part of its evolution. Any observer present to witness this occurrence would see a massive increase in the speed of the solar wind, but not enough to destroy a planet completely. However, the star's loss of mass could send the orbits of the surviving planets into chaos, causing some to collide, others to be ejected from the Solar System, and still others to be torn apart by tidal interactions. Afterwards, all that will remain of the Sun is

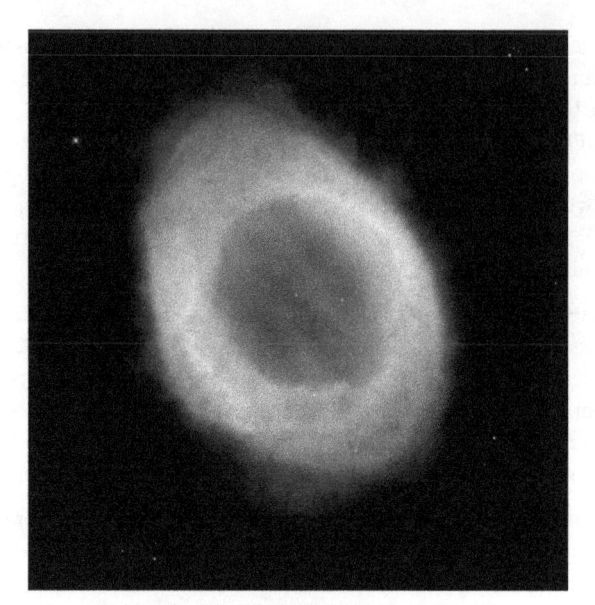

Figure 58: *The Ring nebula, a planetary nebula similar to what the Sun will become*

a white dwarf, an extraordinarily dense object, 54% its original mass but only the size of the Earth. Initially, this white dwarf may be 100 times as luminous as the Sun is now. It will consist entirely of degenerate carbon and oxygen, but will never reach temperatures hot enough to fuse these elements. Thus the white dwarf Sun will gradually cool, growing dimmer and dimmer.

As the Sun dies, its gravitational pull on the orbiting bodies such as planets, comets and asteroids will weaken due to its mass loss. All remaining planets' orbits will expand; if Venus, Earth, and Mars still exist, their orbits will lie roughly at 1.4 AU (210,000,000 km), 1.9 AU (280,000,000 km), and 2.8 AU (420,000,000 km). They and the other remaining planets will become dark, frigid hulks, completely devoid of any form of life. They will continue to orbit their star, their speed slowed due to their increased distance from the Sun and the Sun's reduced gravity. Two billion years later, when the Sun has cooled to the 6000–8000K range, the carbon and oxygen in the Sun's core will freeze, with over 90% of its remaining mass assuming a crystalline structure. Eventually, after billions more years, the Sun will finally cease to shine altogether, becoming a black dwarf.

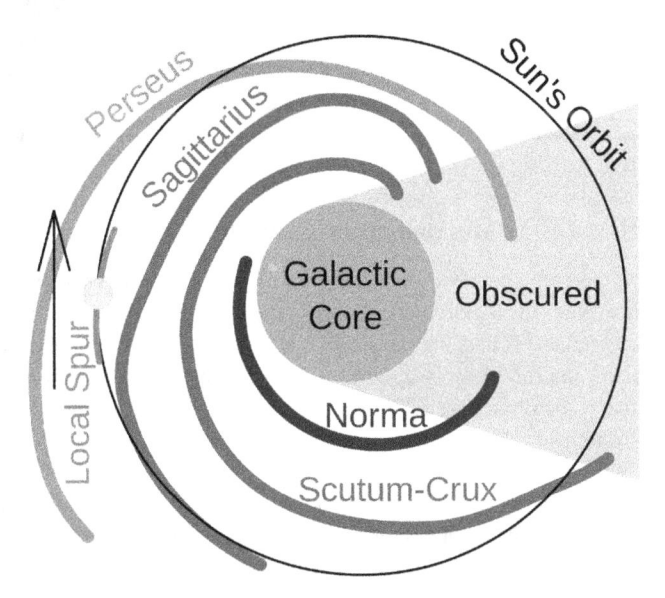

Figure 59: *Location of the Solar System within the Milky Way*

Galactic interaction

The Solar System travels alone through the Milky Way in a circular orbit approximately 30,000 light years from the Galactic Centre. Its speed is about 220 km/s. The period required for the Solar System to complete one revolution around the Galactic Centre, the galactic year, is in the range of 220–250 million years. Since its formation, the Solar System has completed at least 20 such revolutions.

Various scientists have speculated that the Solar System's path through the galaxy is a factor in the periodicity of mass extinctions observed in the Earth's fossil record. One hypothesis supposes that vertical oscillations made by the Sun as it orbits the Galactic Centre cause it to regularly pass through the galactic plane. When the Sun's orbit takes it outside the galactic disc, the influence of the galactic tide is weaker; as it re-enters the galactic disc, as it does every 20–25 million years, it comes under the influence of the far stronger "disc tides", which, according to mathematical models, increase the flux of Oort cloud comets into the Solar System by a factor of 4, leading to a massive increase in the likelihood of a devastating impact.

However, others argue that the Sun is currently close to the galactic plane, and yet the last great extinction event was 15 million years ago. Therefore, the Sun's vertical position cannot alone explain such periodic extinctions, and that extinctions instead occur when the Sun passes through the galaxy's spiral arms. Spiral arms are home not only to larger numbers of molecular clouds, whose gravity may distort the Oort cloud, but also to higher concentrations of bright blue giants, which live for relatively short periods and then explode violently as supernovae.

Galactic collision and planetary disruption

Although the vast majority of galaxies in the Universe are moving away from the Milky Way, the Andromeda Galaxy, the largest member of the Local Group of galaxies, is heading toward it at about 120 km/s. In 4 billion years, Andromeda and the Milky Way will collide, causing both to deform as tidal forces distort their outer arms into vast tidal tails. If this initial disruption occurs, astronomers calculate a 12% chance that the Solar System will be pulled outward into the Milky Way's tidal tail and a 3% chance that it will become gravitationally bound to Andromeda and thus a part of that galaxy. After a further series of glancing blows, during which the likelihood of the Solar System's ejection rises to 30%, the galaxies' supermassive black holes will merge. Eventually, in roughly 6 billion years, the Milky Way and Andromeda will complete their merger into a giant elliptical galaxy. During the merger, if there is enough gas, the increased gravity will force the gas to the centre of the forming elliptical galaxy. This may lead to a short period of intensive star formation called a starburst. In addition, the infalling gas will feed the newly formed black hole, transforming it into an active galactic nucleus. The force of these interactions will likely push the Solar System into the new galaxy's outer halo, leaving it relatively unscathed by the radiation from these collisions.

It is a common misconception that this collision will disrupt the orbits of the planets in the Solar System. Although it is true that the gravity of passing stars can detach planets into interstellar space, distances between stars are so great that the likelihood of the Milky Way–Andromeda collision causing such disruption to any individual star system is negligible. Although the Solar System as a whole could be affected by these events, the Sun and planets are not expected to be disturbed.

However, over time, the cumulative probability of a chance encounter with a star increases, and disruption of the planets becomes all but inevitable. Assuming that the Big Crunch or Big Rip scenarios for the end of the Universe do not occur, calculations suggest that the gravity of passing stars will have completely stripped the dead Sun of its remaining planets within 1 quadrillion (10^{15}) years. This point marks the end of the Solar System. Although the Sun and planets may survive, the Solar System, in any meaningful sense, will cease to exist.

Chronology

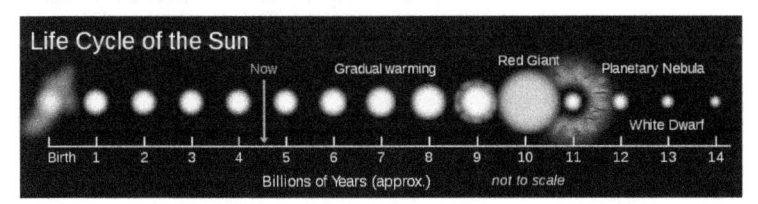

Life Cycle of the Sun

The time frame of the Solar System's formation has been determined using radiometric dating. Scientists estimate that the Solar System is 4.6 billion years old. The oldest known mineral grains on Earth are approximately 4.4 billion years old. Rocks this old are rare, as Earth's surface is constantly being reshaped by erosion, volcanism, and plate tectonics. To estimate the age of the Solar System, scientists use meteorites, which were formed during the early condensation of the solar nebula. Almost all meteorites (see the Canyon Diablo meteorite) are found to have an age of 4.6 billion years, suggesting that the Solar System must be at least this old.

Studies of discs around other stars have also done much to establish a time frame for Solar System formation. Stars between one and three million years old have discs rich in gas, whereas discs around stars more than 10 million years old have little to no gas, suggesting that giant planets within them have ceased forming.

Timeline of Solar System evolution

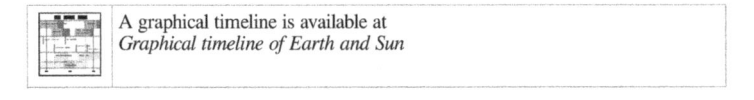

A graphical timeline is available at
Graphical timeline of Earth and Sun

Note: All dates and times in this chronology are approximate and should be taken as an order of magnitude indicator only.

Chronology of the formation and evolution of the Solar System

Phase	Time since formation of the Sun	Time from present (approximate)	Event
Pre-Solar System	Billions of years before the formation of the Solar System	Over 4.6 billion years ago (bya)	Previous generations of stars live and die, injecting heavy elements into the interstellar medium out of which the Solar System formed.

	~ 50 million years before formation of the Solar System	4.6 bya	If the Solar System formed in an Orion nebula-like star-forming region, the most massive stars are formed, live their lives, die, and explode in supernova. One particular supernova, called the *primal supernova*, possibly triggers the formation of the Solar System.
Formation of Sun	0–100,000 years	4.6 bya	Pre-solar nebula forms and begins to collapse. Sun begins to form.
	100,000 – 50 million years	4.6 bya	Sun is a T Tauri protostar.
	100,000 – 10 million years	4.6 bya	By 10 million years, gas in the protoplanetary disc has been blown away, and outer planet formation is likely complete.
	10 million – 100 million years	4.5–4.6 bya	Terrestrial planets and the Moon form. Giant impacts occur. Water delivered to Earth.
Main sequence	50 million years	4.5 bya	Sun becomes a main-sequence star.
	200 million years	4.4 bya	Oldest known rocks on the Earth formed.
	500 million – 600 million years	4.0–4.1 bya	Resonance in Jupiter and Saturn's orbits moves Neptune out into the Kuiper belt. Late Heavy Bombardment occurs in the inner Solar System.
	800 million years	3.8 bya	Oldest known life on Earth. Oort cloud reaches maximum mass.
	4.6 billion years	**Today**	Sun remains a main-sequence star, continually growing warmer and brighter by ~10% every 1 billion years.
	6 billion years	1.4 billion years in the future	Sun's habitable zone moves outside of the Earth's orbit, possibly shifting onto Mars's orbit.
	7 billion years	2.4 billion years in the future	The Milky Way and Andromeda Galaxy begin to collide. Slight chance the Solar System could be captured by Andromeda before the two galaxies fuse completely.
Post–main sequence	10 billion – 12 billion years	5–7 billion years in the future	Sun starts burning hydrogen in a shell surrounding its core, ending its main sequence life. Sun begins to ascend the red giant branch of the Hertzsprung–Russell diagram, growing dramatically more luminous (by a factor of up to 2,700), larger (by a factor of up to 250 in radius), and cooler (down to 2600 K): Sun is now a red giant. Mercury and possibly Venus and Earth are swallowed. Saturn's moon Titan may become habitable.
	~ 12 billion years	~ 7 billion years in the future	Sun passes through helium-burning horizontal-branch and asymptotic-giant-branch phases, losing a total of ~30% of its mass in all post-main-sequence phases. The asymptotic-giant-branch phase ends with the ejection of a planetary nebula, leaving the core of the Sun behind as a white dwarf.

| Remnant Sun | ~1 quadrillion years (10^{15} years) | ~1 quadrillion years in the future | Sun cools to 5 K. Gravity of passing stars detaches planets from orbits. Solar System ceases to exist. |

Bibliography

- Duncan, Martin J.; Lissauer, Jack J. (1997). "Orbital Stability of the Uranian Satellite System". *Icarus.* **125** (1): 1–12. Bibcode: 1997Icar..125....1D[149]. doi: 10.1006/icar.1996.5568[150].
- Zeilik, Michael A.; Gregory, Stephen A. (1998). *Introductory Astronomy & Astrophysics* (4th ed.). Saunders College Publishing. ISBN 0-03-006228-4.

External links

- 7M animation[151] from skyandtelescope.com[152] showing the early evolution of the outer Solar System.
- QuickTime animation of the future collision between the Milky Way and Andromeda[153]
- How the Sun Will Die: And What Happens to Earth[154] (Video at Space.com)

<indicator name="featured-star"> ⭐ </indicator>

Nice model

The **Nice** (/'niːs/) **model** is a scenario for the dynamical evolution of the Solar System. It is named for the location of the Observatoire de la Côte d'Azur, where it was initially developed, in Nice, France. It proposes the migration of the giant planets from an initial compact configuration into their present positions, long after the dissipation of the initial protoplanetary disk. In this way, it differs from earlier models of the Solar System's formation. This planetary migration is used in dynamical simulations of the Solar System to explain historical events including the Late Heavy Bombardment of the inner Solar System, the formation of the Oort cloud, and the existence of populations of small Solar System bodies including the Kuiper belt, the Neptune and Jupiter trojans, and the numerous resonant trans-Neptunian objects dominated by Neptune. Its success at reproducing many of the observed features of the Solar System means that it is widely accepted as the current most realistic model of the Solar System's early evolution, although it is not universally favoured among planetary scientists. Later research revealed a number of differences between the

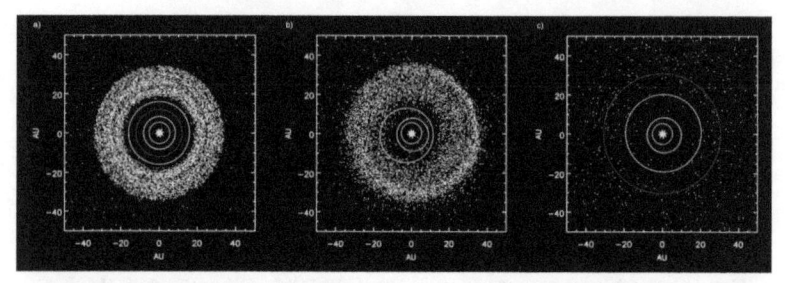

Figure 60: *Simulation showing the outer planets and planetesimal belt: a) early configuration, before Jupiter and Saturn reach a 2:1 resonance; b) scattering of planetesimals into the inner Solar System after the orbital shift of Neptune (dark blue) and Uranus (light blue); c) after ejection of planetesimals by planets.*

original Nice model's predictions and observations of the current Solar System, for example the orbits of the terrestrial planets and the asteroids, leading to its modification.

Description

The original core of the Nice model is a triplet of papers published in the general science journal *Nature* in 2005 by an international collaboration of scientists: Rodney Gomes, Hal Levison, Alessandro Morbidelli, and Kleomenis Tsiganis. In these publications, the four authors proposed that after the dissipation of the gas and dust of the primordial Solar System disk, the four giant planets (Jupiter, Saturn, Uranus, and Neptune) were originally found on near-circular orbits between ~ 5.5 and ~ 17 astronomical units (AU), much more closely spaced and compact than in the present. A large, dense disk of small rock and ice planetesimals, their total about 35 Earth masses, extended from the orbit of the outermost giant planet to some 35 AU.

Scientists understand so little about the formation of Uranus and Neptune that Levison states, "...the possibilities concerning the formation of Uranus and Neptune are almost endless." However, it is suggested that this planetary system evolved in the following manner. Planetesimals at the disk's inner edge occasionally pass through gravitational encounters with the outermost giant planet, which change the planetesimals' orbits. The planets scatter the majority of the small icy bodies that they encounter inward, exchanging angular momentum with the scattered objects so that the planets move outwards in response, preserving the angular momentum of the system. These planetesimals then similarly scatter off the next planet they encounter, successively moving the orbits of Uranus, Neptune, and Saturn outwards. Despite the minute movement each exchange of momentum can produce, cumulatively these planetesimal encounters shift (migrate) the orbits of the planets by significant amounts. This process continues until the planetesimals interact with the innermost and most massive giant planet, Jupiter, whose immense gravity sends them into highly elliptical orbits or even ejects them outright from the Solar System. This, in contrast, causes Jupiter to move slightly inward.

The low rate of orbital encounters governs the rate at which planetesimals are lost from the disk, and the corresponding rate of migration. After several hundreds of millions of years of slow, gradual migration, Jupiter and Saturn, the two inmost giant planets, cross their mutual 1:2 mean-motion resonance. This resonance increases their orbital eccentricities, destabilizing the entire planetary system. The arrangement of the giant planets alters quickly and dramatically. Jupiter shifts Saturn out towards its present position, and this relocation causes mutual gravitational encounters between Saturn and the two ice giants, which propel Neptune and Uranus onto much more eccentric orbits. These ice giants then plough into the planetesimal disk, scattering tens of thousands of planetesimals from their formerly stable orbits in the outer Solar System. This disruption almost entirely scatters the primordial disk, removing 99% of its mass, a scenario which explains the modern-day absence of a dense trans-Neptunian population. Some of the planetesimals are thrown into the inner Solar System, producing a sudden influx of impacts on the terrestrial planets: the Late Heavy Bombardment.

Eventually, the giant planets reach their current orbital semi-major axes, and dynamical friction with the remaining planetesimal disc damps their eccentricities and makes the orbits of Uranus and Neptune circular again.

In some 50% of the initial models of Tsiganis and colleagues, Neptune and Uranus also exchange places. An exchange of Uranus and Neptune would be consistent with models of their formation in a disk that had a surface density that declined with distance from the Sun, which predicts that the masses of the planets should also decline with distance from the Sun.

Solar System features

Running dynamical models of the Solar System with different initial conditions for the simulated length of the history of the Solar System will produce the various populations of objects within the Solar System. As the initial conditions of the model are allowed to vary, each population will be more or less numerous, and will have particular orbital properties. Proving a model of the evolution of the early Solar System is difficult, since the evolution cannot be directly observed. However, the success of any dynamical model can be judged

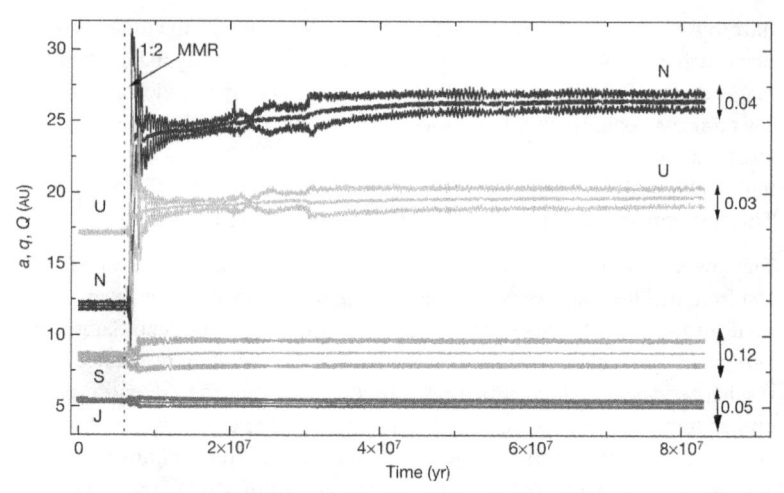

Figure 61: *Example Nice Model simulation of the migration of the solar distance of the four giant planets.*

by comparing the population predictions from the simulations to astronomical observations of these populations. At the present time, computer models of the Solar System that are begun with the initial conditions of the Nice scenario best match many aspects of the observed Solar System.

The Late Heavy Bombardment

The crater record on the Moon and on the terrestrial planets is part of the main evidence for the Late Heavy Bombardment (LHB): an intensification in the number of impactors, at about 600 million years after the Solar System's formation. In the Nice model icy planetesimals are scattered onto planet-crossing orbits when the outer disc is disrupted by Uranus and Neptune causing a sharp spike of impacts by icy objects. The migration of outer planets also causes mean-motion and secular resonances to sweep through the inner Solar System. In the asteroid belt these excite the eccentricities of the asteroids driving them onto orbits that intersect those of the terrestrial planets causing a more extended period of impacts by stony objects and removing roughly 90% of its mass. The number of planetesimals that would reach the Moon is consistent with the crater record from the LHB. However, the orbital distribution of the remaining asteroids does not match observations. In the outer Solar System the impacts onto Jupiter's moons are sufficient to trigger Ganymede's differentiation but not Callisto's. The impacts of icy planetesimals onto Saturn's inner moons are excessive, however, resulting in the vaporization of their ice.

Trojans and the asteroid belt

After Jupiter and Saturn cross the 2:1 resonance their combined gravitational influence destabilizes the Trojan co-orbital region allowing existing Trojan groups in the L_4 and L_5 Lagrange points of Jupiter and Neptune to escape and new objects from the outer planetesimal disk to be captured. Objects in the trojan co-orbital region undergo libration, drifting cyclically relative to the L_4 and L_5 points. When Jupiter and Saturn are near but not in resonance the location where Jupiter passes Saturn relative to their perihelia circulates slowly. If the period of this circulation falls into resonance with the period that the trojans librate the range of their librations can increase until they escape. When this occurs the trojan co-orbital region is "dynamically open" and objects can both escape and enter the region. Primordial trojans escape and a fraction of the numerous objects from the disrupted planetesimal disk temporarily inhabit it. Later when Jupiter and Saturn orbits are farther apart the Trojan region becomes "dynamically closed", and the planetesimals in the trojan region are captured, with many remaining today. The captured trojans have a wide range of inclinations, which had not previously been understood, due to their repeated encounters with the giant planets. The libration angle and eccentricity of the simulated population also matches observations of the orbits of the Jupiter trojans. This mechanism of the Nice model similarly generates the Neptune trojans.

A large number of planetesimals would have also been captured in Jupiter's mean motion resonances as Jupiter migrated inward. Those that remained in a 3:2 resonance with Jupiter form the Hilda family. The eccentricity of other objects declined while they were in a resonance and escaped onto stable orbits in the outer asteroid belt, at distances greater than 2.6 AU as the resonances moved inward. These captured objects would then have undergone collisional erosion, grinding the population away into smaller fragments that can then be acted on by the Yarkovsky effect, causing small objects to drift into unstable resonances, and Poynting–Robertson drag causing smaller grains to drift toward the sun. These processes remove more than 90% of the origin mass implanted into the asteroid belt according to Bottke and colleagues. The size frequency distribution of this simulated population following this erosion are in excellent agreement with observations. This suggests that the Jupiter Trojans, Hildas, and some of the outer asteroid belt, all spectral D-type asteroids, are the remnant planetesimals from this capture and erosion process. It has also been suggested that the dwarf planet Ceres was captured via this process. A few D-type asteroids have been recently discovered with semi-major axes less than 2.5 AU, closer than those that would be captured in the original Nice model.

Outer-system satellites

Any original populations of irregular satellites captured by traditional mechanisms, such as drag or impacts from the accretion disks,[155] would be lost during the encounters between the planets at the time of global system instability. In the Nice model, the outer planets encounter large numbers of planetesimals after Uranus and Neptune enter and disrupt the planetesimal disk. A fraction of these planetesimals are captured by these planets via three-way interactions during encounters between planets. The probability for any planetesimal to be captured by an ice giant is relatively high, a few 10^{-7}. These new satellites could be captured at almost any angle, so unlike the regular satellites of Saturn, Uranus, and Neptune, they do not necessarily orbit in the planets' equatorial planes. Some irregulars may have even been exchanged between planets. The resulting irregular orbits match well with the observed populations' semimajor axes, inclinations, and eccentricities. Subsequent collisions between these captured satellites may have created the suspected collisional families seen today. These collisions are also required to erode the population to the present size distribution.

Triton, the largest moon of Neptune, can be explained if it was captured in a three-body interaction involving the disruption of a binary planetoid. Such binary disruption would be more likely if Triton was the smaller member of the binary. However, Triton's capture would be more likely in the early Solar System when the gas disk would damp relative velocities, and binary exchange reactions would not in general have supplied the large number of small irregulars.

There were not enough interactions between Jupiter and the other planets to explain Jupiter's retinue of irregulars in the initial Nice model simulations that reproduced other aspects of the outer Solar System. This suggests either that a second mechanism was at work for that planet, or that the early simulations did not reproduce the evolution of the giant planets orbits.

Formation of the Kuiper belt

The migration of the outer planets is also necessary to account for the existence and properties of the Solar System's outermost regions. Originally, the Kuiper belt was much denser and closer to the Sun, with an outer edge at approximately 30 AU. Its inner edge would have been just beyond the orbits of Uranus and Neptune, which were in turn far closer to the Sun when they formed (most likely in the range of 15–20 AU), and in opposite locations, with Uranus farther from the Sun than Neptune.

Gravitational encounters between the planets scatter Neptune outward into the planetesimal disk with a semi-major axis of $\sim$28 AU and an eccentricity as high as 0.4. Neptune's high eccentricity causes its mean-motion resonances to overlap and orbits in the region between Neptune and its 2:1 mean motion resonances to become chaotic. The orbits of objects between Neptune and the edge of the planetesimal disk at this time can evolve outward onto stable low-eccentricity orbits within this region. When Neptune's eccentricity is damped by dynamical friction they become trapped on these orbits. These objects form a dynamically-cold belt, since their inclinations remain small during the short time they interact with Neptune. Later, as Neptune migrates outward on a low eccentricity orbit, objects that have been scattered outward are captured into its resonances and can have their eccentricities decline and their inclinations increase due to the Kozai mechanism, allowing them to escape onto stable higher-inclination orbits. Other objects remain captured in resonance, forming the plutinos and other resonant populations. These two population are dynamically hot, with higher inclinations and eccentricities; due to their being scattered outward and the longer period these objects interact with Neptune.

This evolution of Neptune's orbit produces both resonant and non-resonant populations, an outer edge at Neptune's 2:1 resonance, and a small mass relative to the original planetesimal disk. The excess of low-inclination plutinos in other models is avoided due to Neptune being scattered outward, leaving its 3:2 resonance beyond the original edge of the planetesimal disk. The differing initial locations, with the cold classical objects originating primarily from the outer disk, and capture processes, offer explanations for the bi-modal inclination distribution and its correlation with compositions. However, this evolution of Neptune's orbit fails to account for some of the characteristics of the orbital distribution. It predicts a greater average eccentricity in classical Kuiper belt object orbits than is observed (0.10–0.13 versus 0.07) and it does not produce enough higher-inclination objects. It also cannot explain the apparent complete absence of gray objects in the cold population, although it has been suggested that color differences arise in part from surface evolution processes rather than entirely from differences in primordial composition.

The shortage of the lowest-eccentricity objects predicted in the Nice model may indicate that the cold population formed in situ. In addition to their differing orbits the hot and cold populations have but differing colors. The cold population is markedly redder than the hot, suggesting it has a different composition and formed in a different region. The cold population also includes a large number of binary objects with loosely bound orbits that would be unlikely to survive close encounter with Neptune. If the cold population formed at its current location, preserving it would require that Neptune's eccentricity remained small, or that its perihelion precessed rapidly due to a strong interaction between it and Uranus.

Scattered disc and Oort cloud

Objects scattered outward by Neptune onto orbits with semi-major axis greater than 50 AU can be captured in resonances forming the resonant population of the scattered disc, or if their eccentricities are reduced while in resonance they can escape from the resonance onto stable orbits in the scattered disc while Neptune is migrating. When Neptune's eccentricity is large its aphelion can reach well beyond its current orbit. Objects that attain perihelia close to or larger than Neptune's at this time can become detached from Neptune when its eccentricity is damped reducing its aphelion, leaving them on stable orbits in the scattered disc.

Objects scattered outward by Uranus and Neptune onto larger orbits (roughly 5,000 AU) can have their perihelion raised by the galactic tide detaching them from the influence of the planets forming the inner Oort cloud with moderate inclinations. Others that reach even larger orbits can be perturbed by nearby stars forming the outer Oort cloud with isotropic inclinations. Objects scattered by Jupiter and Saturn are typically ejected from the Solar System Several percent of the initial planetesimal disc can be deposited in these reservoirs.

Modifications

The Nice model has undergone a number of modifications since its initial publication as the understanding of the formation of the Solar System has advanced and significant differences between its predictions and observations have been identified. Hydrodynamical models of the early Solar System indicate that the orbits of the giant planet converge resulting in their capture into resonances. During the slow approach of Jupiter and Saturn to the 2:1 resonance, Mars can be captured in a secular resonance, exciting its eccentricity to a level that destabilizes the inner Solar System. The eccentricities of the other terrestrial planets can also be excited beyond current levels by sweeping secular resonances after the instability. The orbital distribution of the asteroid belt is also left with an excess of high inclination objects due to secular resonances exciting inclinations and removing low inclination objects. Other differences between predictions and observations included the capture of few irregular satellites by Jupiter, the vaporization of the ice from Saturn's inner moons, a shortage of high inclination objects captured in the Kuiper belt, and the recent discovery of D-type asteroids in the inner asteroid belt.

The first modifications to the Nice model were the initial positions of the giant planets. Investigations of the behavior of planets orbiting in a gas disk using hydrodynamical models reveal that the giant planets would migrate toward the Sun. If the migration continued it would have resulted in Jupiter orbiting close to the Sun like recently discovered exoplanets known as hot Jupiters. Saturn's capture in a resonance with Jupiter prevents this, however, and the later capture of the other planets results in a quadruple resonant configuration with Jupiter and Saturn in their 3:2 resonance. A late instability beginning from this configuration is possible if the outer disk contains Pluto-massed objects. The gravitational stirring of the outer planetesimal disk by these Pluto-massed objects increases their eccentricities and also results in the inward migration of the giant planets. The quadruple resonance of the giant planets is broken when secular resonances are crossed during the inward migration. A late instability similar to the original Nice model then follows. Unlike the original Nice model the timing of this instability is not sensitive to the distance between the outer planet and the planetesimal disk. The combination of resonant planetary orbits and the late instability triggered by these long distant interactions has been referred to as the Nice 2 model.

The second modification was the requirement that one of the ice giants encounters Jupiter, causing its semi-major axis to jump. In this jumping-Jupiter scenario an ice giant encounters Saturn and is scattered inward onto a Jupiter-crossing orbit, causing Saturn's orbit to expand; then encounters Jupiter and is scattered outward, causing Jupiter's orbit to shrink. This results in a step-wise separation of Jupiter's and Saturn's orbits instead of a smooth divergent migration. The step-wise separation of the orbits of Jupiter and Saturn avoids the slow sweeping of secular resonances across the inner solar System that resulted in the excitation of the eccentricities of the terrestrial planets and an asteroid belt with an excessive ratio of high- to low-inclination objects The encounters between the ice giant and Jupiter in this model allow Jupiter to acquire its own irregular satellites. Jupiter trojans are also captured following these encounters when Jupiter's semi-major axis jumps and, if the ice giant passes through one of the libration points scattering trojans, one population is depleted relative to the other. The faster traverse of the secular resonances across the asteroid belt limits the loss of asteroids from its core. Most of the rocky impactors of the Late Heavy Bombardment instead originate from an inner extension that is disrupted when the giant planets reach their current positions, with a remnant remaining as the Hungaria asteroids. Some D-type asteroids are embedded in inner asteroid belt, within 2.5 AU, during encounters with the ice giant when it is crossing the asteroid belt.

Five-planet Nice model

The frequent ejection of the ice giant encountering Jupiter has led David Nesvorný and others to hypothesize an early Solar System with five giant planets, one of which was ejected during the instability. This five-planet Nice model begins with the giant planets in a 3:2, 3:2, 2:1, 3:2 resonant chain with a planetesimal disk orbiting beyond them. Following the breaking of the resonant chain Neptune first migrates outward into the planetesimal disk reaching 28 AU before encounters between planets begin. This initial migration reduces the mass of the outer disk enabling Jupiter's eccentricity to be preserved and produces a Kuiper belt with an inclination distribution that matches observations if 20 Earth-masses remained in the planetesimal disk when that migration began. Neptune's eccentricity can remain small during the instability since it only encounters the ejected ice giant, allowing an in situ cold-classical belt to be preserved. The lower mass planetesimal belt in combination with the excitation of inclinations and eccentricities by the Pluto-massed objects also significantly reduce the loss of ice by Saturn's inner moons. The combination of a late breaking of the resonance chain and a migration of Neptune to 28 AU before the instability is unlikely with the Nice 2 model. This gap may be bridged by a slow dust-driven migration over several million years following an early escape from resonance. A recent study found that the five-planet Nice model has a statistically small likelihood of reproducing the orbits of the terrestrial planets. Although this implies that the instability occurred before the formation of the terrestrial planets and could not be the source of the Late Heavy Bombardment, the advantage of an early instability is reduced by the sizable jumps in the semi-major axis of Jupiter and Saturn required to preserve the asteroid belt.

External links

Wikimedia Commons has media related to *Nice Model*.

- Animation of the Nice model[156]
- Solving solar system quandaries is simple: Just flip-flop the position of Uranus and Neptune[157]

Moons

Moons of Neptune

<indicator name="featured-star"> ⭐ </indicator>

Neptune has 14 known moons, which are named for minor water deities in Greek mythology.[158] By far the largest of them is Triton, discovered by William Lassell on October 10, 1846, 17 days after the discovery of Neptune itself; over a century passed before the discovery of the second natural satellite, Nereid. Neptune's outermost moon Neso, which has an orbital period of about 26 Julian years, orbits further from its planet than any other moon in the Solar System.[159]

Triton is unique among moons of planetary mass in that its orbit is retrograde to Neptune's rotation and inclined relative to Neptune's equator, which suggests that it did not form in orbit around Neptune but was instead gravitationally captured by it. The next-largest irregular satellite in the Solar System, Saturn's moon Phoebe, has only 0.03% of Triton's mass. The capture of Triton, probably occurring some time after Neptune formed a satellite system, was a catastrophic event for Neptune's original satellites, disrupting their orbits so that they collided to form a rubble disc. Triton is massive enough to have achieved hydrostatic equilibrium and to retain a thin atmosphere capable of forming clouds and hazes.

Inward of Triton are seven small regular satellites, all of which have prograde orbits in planes that lie close to Neptune's equatorial plane; some of these orbit among Neptune's rings. The largest of them is Proteus. They were re-accreted from the rubble disc generated after Triton's capture after the Tritonian orbit became circular. Neptune also has six more outer irregular satellites other than Triton, including Nereid, whose orbits are much farther from Neptune and at high inclination: three of these have prograde orbits, while the remainder have retrograde orbits. In particular, Nereid has an unusually close and eccentric orbit for an irregular satellite, suggesting that it may have once been a regular

Figure 62: *Neptune (top) and Triton (bottom), three days after the Voyager 2 flyby in 1989*

satellite that was significantly perturbed to its current position when Triton was captured. The two outermost Neptunian irregular satellites, Psamathe and Neso, have the largest orbits of any natural satellites discovered in the Solar System to date.

History

Discovery

Triton was discovered by William Lassell in 1846, just seventeen days after the discovery of Neptune. Nereid was discovered by Gerard P. Kuiper in 1949. The third moon, later named Larissa, was first observed by Harold J. Reitsema, William B. Hubbard, Larry A. Lebofsky and David J. Tholen on May 24, 1981. The astronomers were observing a star's close approach to Neptune, looking for rings similar to those discovered around Uranus four years earlier. If rings were present, the star's luminosity would decrease slightly just before the planet's closest approach. The star's luminosity dipped only for several seconds, which meant that it was due to a moon rather than a ring.

No further moons were found until *Voyager 2* flew by Neptune in 1989. *Voyager 2* rediscovered Larissa and discovered five inner moons: Naiad, Thalassa,

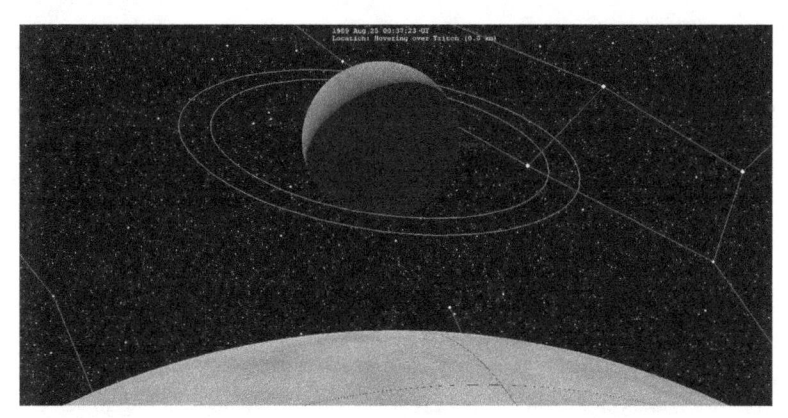

Figure 63: *Simulated view of Neptune in the hypothetical sky of Triton*

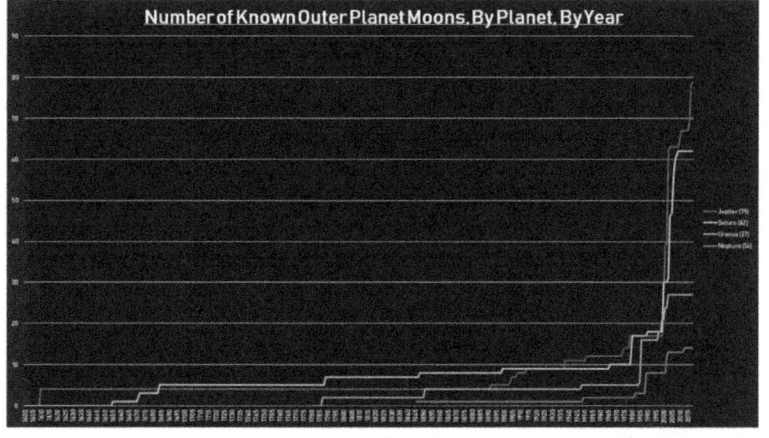

Figure 64: *The number of moons known for each of the four outer planets up to July 2018. Neptune currently has 14 known satellites.*

Despina, Galatea and Proteus. In 2001 two surveys using large ground-based telescopes found five additional outer moons, bringing the total to thirteen. Follow-up surveys by two teams in 2002 and 2003 respectively re-observed all five of these moons, which are Halimede, Sao, Psamathe, Laomedeia, and Neso. A sixth candidate moon was also found in the 2002 survey and was lost thereafter: it may have been a centaur instead of a satellite, although its small amount of motion relative to Neptune over a month suggests that it was indeed a satellite. It was estimated to have a diameter of 33 km and to have been about 25.1 million km (0.168 AU) from Neptune when it was found.

Names

Triton did not have an official name until the twentieth century. The name "Triton" was suggested by Camille Flammarion in his 1880 book *Astronomie Populaire*, but it did not come into common use until at least the 1930s. Until this time it was usually simply known as "the satellite of Neptune". Other moons of Neptune are also named for Greek and Roman water gods, in keeping with Neptune's position as god of the sea: either from Greek mythology, usually children of Poseidon, the Greek Neptune (Triton, Proteus, Despina, Thalassa); lovers of Poseidon (Larissa); classes of minor Greek water deities (Naiad, Nereid); or specific Nereids (Halimede, Galatea, Neso, Sao, Laomedeia, Psamathe). For the "normal" irregular satellites, the convention is to use names ending in "a" for prograde satellites, names ending in "e" for retrograde satellites, and names ending in "o" for exceptionally inclined satellites, exactly like the convention for the moons of Jupiter. Two asteroids share the same names as moons of Neptune: 74 Galatea and 1162 Larissa.

Characteristics

The moons of Neptune can be divided into two groups: regular and irregular. The first group includes the seven inner moons, which follow circular prograde orbits lying in the equatorial plane of Neptune. The second group consists of all seven other moons including Triton. They generally follow inclined eccentric and often retrograde orbits far from Neptune; the only exception is Triton, which orbits close to the planet following a circular orbit, though retrograde and inclined.

Regular moons

In order of distance from Neptune, the regular moons are Naiad, Thalassa, Despina, Galatea, Larissa, S/2004 N 1, and Proteus. All but the outer two are within Neptune-synchronous orbit (Neptune's rotational period is 0.6713 day) and thus are being tidally decelerated. Naiad, the closest regular moon, is also the second smallest among the inner moons (following the discovery of S/2004 N 1), whereas Proteus is the largest regular moon and the second largest moon of Neptune.

The inner moons are closely associated with Neptune's rings. The two innermost satellites, Naiad and Thalassa, orbit between the Galle and LeVerrier rings. Despina may be a shepherd moon of the LeVerrier ring, because its orbit lies just inside this ring. The next moon, Galatea, orbits just inside the most prominent of Neptune's rings, the Adams ring. This ring is very narrow, with a width not exceeding 50 km, and has five embedded bright arcs. The gravity

Moons of Neptune

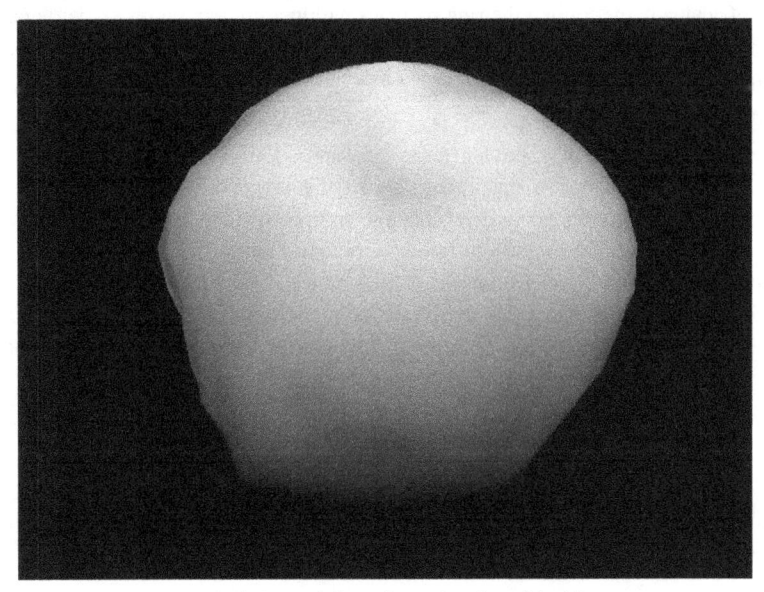

Figure 65: *Animated three-dimensional model of Proteus*

Figure 66: *A time-lapse video depicting orbits of Neptune's moons: Triton, Proteus, Larissa, Galatea and Despina.*

of Galatea helps confine the ring particles within a limited region in the radial direction, maintaining the narrow ring. Various resonances between the ring particles and Galatea may also have a role in maintaining the arcs.

Only the two largest regular moons have been imaged with a resolution sufficient to discern their shapes and surface features. Larissa, about 200 km in diameter, is elongated. Proteus is not significantly elongated, but not fully spherical either: it resembles an irregular polyhedron, with several flat or slightly concave facets 150 to 250 km in diameter. At about 400 km in diameter, it is larger than the Saturnian moon Mimas, which is fully ellipsoidal. This difference may be due to a past collisional disruption of Proteus. The surface of Proteus is heavily cratered and shows a number of linear features. Its largest crater, Pharos, is more than 150 km in diameter.

All of Neptune's inner moons are dark objects: their geometric albedo ranges from 7 to 10%. Their spectra indicate that they are made from water ice contaminated by some very dark material, probably complex organic compounds. In this respect, the inner Neptunian moons are similar to the inner Uranian moons.

Irregular moons

In order of their distance from the planet, the irregular moons are Triton, Nereid, Halimede, Sao, Laomedeia, Psamathe, and Neso, a group that includes both prograde and retrograde objects. The five outermost moons are similar to the irregular moons of other giant planets, and are thought to have been gravitationally captured by Neptune, unlike the regular satellites, which probably formed *in situ*.

Triton and Nereid are unusual irregular satellites and are thus treated separately from the other five irregular Neptunian moons, which are more like the outer irregular satellites of the other outer planets. Firstly, they are the largest two known irregular moons in the Solar System, with Triton being almost an order of magnitude larger than all other known irregular moons. Secondly, they both have atypically small semi-major axes, with Triton's being over an order of magnitude smaller than those of all other known irregular moons. Thirdly, they both have unusual orbital eccentricities: Nereid has one of the most eccentric orbits of any known irregular satellite, and Triton's orbit is a nearly perfect circle. Finally, Nereid also has the lowest inclination of any known irregular satellite.

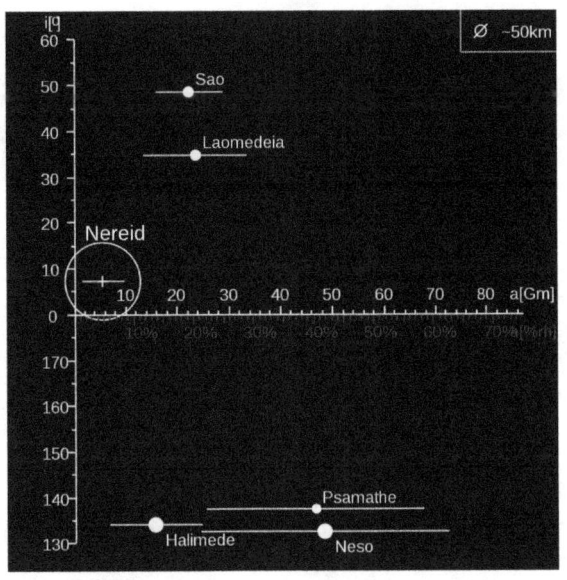

Figure 67: *The diagram illustrates the orbits of Neptune's irregular moons excluding Triton. The eccentricity is represented by the yellow segments extending from the pericenter to apocenter with the inclination represented on Y axis. The moons above the X axis are prograde, those beneath are retrograde. The X axis is labeled in Gm and the fraction of the Hill sphere's radius.*

Triton

Triton follows a retrograde and quasi-circular orbit, and is thought to be a gravitationally captured satellite. It was the second moon in the Solar System that was discovered to have a substantial atmosphere, which is primarily nitrogen with small amounts of methane and carbon monoxide. The pressure on Triton's surface is about 14 µbar. In 1989 the *Voyager 2* spacecraft observed what appeared to be clouds and hazes in this thin atmosphere. Triton is one of the coldest bodies in the Solar System, with a surface temperature of about 38 K (–235.2 °C). Its surface is covered by nitrogen, methane, carbon dioxide and water ices and has a high geometric albedo of more than 70%. The Bond albedo is even higher, reaching up to 90%.[160] Surface features include the large southern polar cap, older cratered planes cross-cut by graben and scarps, as well as youthful features probably formed by endogenic processes like cryovolcanism. *Voyager 2* observations revealed a number of active geysers within the polar cap heated by the Sun, which eject plumes to the height of up to 8 km. Triton has a relatively high density of about 2 g/cm^3 indicating that rocks constitute about two thirds of its mass, and ices (mainly water ice) the remaining

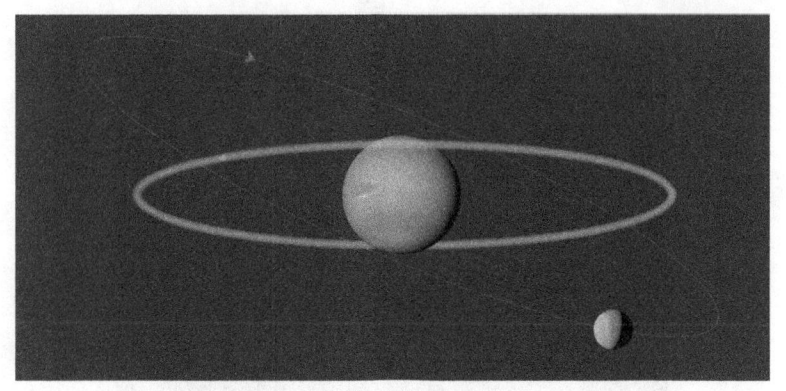

Figure 68: *The orbit of Triton (red) is different from most moons' orbit (green) in the orbit's direction, and the orbit is tilted –23°.*

one third. There may be a layer of liquid water deep inside Triton, forming a subterranean ocean. Because of its retrograde orbit and relative proximity to Neptune (closer than the Moon is to Earth), tidal deceleration is causing Triton to spiral inward, which will lead to its destruction in about 3.6 billion years.

Nereid

Nereid is the third-largest moon of Neptune. It has a prograde but very eccentric orbit and is believed to be a former regular satellite that was scattered to its current orbit through gravitational interactions during Triton's capture. Water ice has been spectroscopically detected on its surface. Early measurements of Nereid showed large, irregular variations in its visible magnitude, which were speculated to be caused by forced precession or chaotic rotation combined with an elongated shape and bright or dark spots on the surface. This was disproved in 2016, when observations from the Kepler space telescope showed only minor variations. Thermal modeling based on infrared observations from the Spitzer and Herschel space telescopes suggest that Nereid is only moderately elongated which disfavours forced precession of the rotation. The thermal model also indicates that the surface roughness of Nereid is very high, likely similar to the Saturnian moon Hyperion.

Normal irregular moons

Among the remaining irregular moons, Sao and Laomedeia follow prograde orbits, whereas Halimede, Psamathe and Neso follow retrograde orbits. Given the similarity of their orbits, it was suggested that Neso and Psamathe could have a common origin in the break-up of a larger moon. Psamathe and Neso have the largest orbits of any natural satellites discovered in the Solar system

■ Triton

▪ all other moons

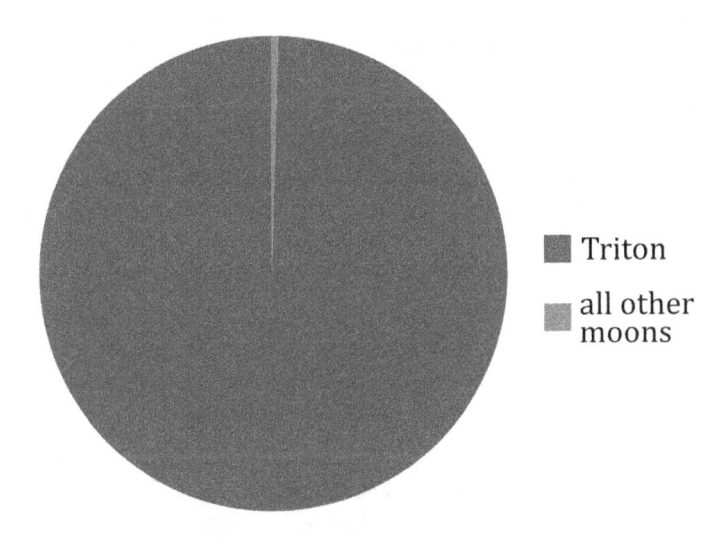

Figure 69: *The relative masses of the Neptunian moons*

to date. They take 25 years to orbit Neptune at an average of 125 times the distance between Earth and the Moon. Neptune has the largest Hill sphere in the Solar System, owing primarily to its large distance from the Sun; this allows it to retain control of such distant moons. Nevertheless, Jupiter's S/2003 J 2 orbits at the greatest percentage of the primary's Hill radius of all the moons in the Solar System on average, and the Jovian moons in the Carme and Pasiphae groups orbit at a greater percentage of their primary's Hill radius than Psamathe and Neso.

Formation

The mass distribution of the Neptunian moons is the most lopsided of the satellite systems of the giant planets in the Solar System. One moon, Triton, makes up nearly all of the mass of the system, with all other moons together comprising only one third of one percent. This is similar to the moon system of Saturn, where Titan makes up more than 95% of the total mass, but is different from the more balanced systems of Jupiter and Uranus. The reason for the lopsidedness of the present Neptunian system is that Triton was captured well after the formation of Neptune's original satellite system, much of which would have been destroyed in the process of capture.

Triton's orbit upon capture would have been highly eccentric, and would have caused chaotic perturbations in the orbits of the original inner Neptunian satellites, causing them to collide and reduce to a disc of rubble. This means it is likely that Neptune's present inner satellites are not the original bodies that formed with Neptune. Only after Triton's orbit became circularised could some of the rubble re-accrete into the present-day regular moons. This great perturbation may possibly be the reason why the satellite system of Neptune does not follow the 10,000:1 ratio of mass between the parent planet and all its moons seen in the satellite systems of all the other giant planets.

The mechanism of Triton's capture has been the subject of several theories over the years. One of them postulates that Triton was captured in a three-body encounter. In this scenario, Triton is the surviving member of a binary Kuiper belt object[161] disrupted by its encounter with Neptune.

Numerical simulations show that there is a 0.41 probability that the moon Halimede collided with Nereid at some time in the past. Although it is not known whether any collision has taken place, both moons appear to have similar ("grey") colors, implying that Halimede could be a fragment of Nereid.

List

Key

‡	♠
Prograde irregular moons	Retrograde irregular moons

The Neptunian moons are listed here by orbital period, from shortest to longest. Irregular (captured) moons are marked by color. Triton, the only Neptunian moon massive enough for its surface to have collapsed into a spheroid, is emboldened.

Neptunian moons

Order[162]	Label[163]	Name	Pronunciation (key)	Image	Diameter (km)[164]	Mass ($\times 10^{16}$ kg)[165]	Semi-major axis (km)	Orbital period (d)	Orbital inclination (°)[166]	Eccentricity	Discovery year	Discoverer
1	III	Naiad	/ˈneɪæd, -ɑːd/[167], <wbr />ˈnaɪ-, <wbr />		66 (96 × 60 × 52)	≈ 19	48227	0.294	4.691	0.0003	1989	Voyager Science Team
2	IV	Thalassa	/θəˈlæsə/		82 (108 × 100 × 52)	≈ 35	50074	0.311	0.135	0.0002	1989	Voyager Science Team
3	V	Despina	/dɪsˈpiːnə/		150 (180 × 148 × 128)	≈ 210	52526	0.335	0.068	0.0002	1989	Voyager Science Team
4	VI	Galatea	/ˌɡæləˈtiːə/		176 (204 × 184 × 144)	≈ 375	61953	0.429	0.034	0.0001	1989	Voyager Science Team
5	VII	Larissa	/ləˈrɪsə/		194 (216 × 204 × 168)	≈ 495	73548	0.555	0.205	0.0014	1981	Reitsema et al.
6	—	S/2004 N 1			≈ 16–20	≈ 0.5 ± 0.4	105300 ± 50	0.936	0.000	0.0000	2013	Showalter et al.

7	VIII	Proteus	/ˈprouti:əs/		420 (436 × 416 × 402)	≈5035	117646	1.122	0.075	0.0005	1989	Voyager Science Team
8	I	**Triton**♠	/ˈtraɪtən/		2705.2±4.8 (2709 × 2706 × 2705)	2140800±5200	354759	−5.877	156.865	0.0000	1846	Lassell
9	II	Nereid‡	/ˈnɪəriːɪd/		≈ 340 ± 50	≈2700	5513818	360.13	7.090	0.7507	1949	Kuiper
10	IX	Halimede♠	/ˌhælɪˈmiːdiː/		≈ 62	≈ 16	16611000	−1879.08	134.1	0.2646	2002	Holman et al.
11	XI	Sao‡	/ˈseɪoʊ/		≈ 44	≈ 6	22228000	2912.72	49.907	0.1365	2002	Holman et al.
12	XII	Laomedeia‡	/ˌleɪoʊmɪˈdiːə/		≈ 42	≈ 5	23567000	3171.33	34.049	0.3969	2002	Holman et al.
13	X	Psamathe♠	/ˈsæməθiː/		≈ 40	≈ 4	48096000	−9074.30	137.679	0.3809	2003	Sheppard et al.
14	XIII	Neso♠	/ˈniːsoʊ/		≈ 60	≈ 15	49285000	−9740.73	131.265	0.5714	2002	Holman et al.

References
External links

> Wikimedia Commons has media related to *Moons of Neptune*.

- Neptune's Known Satellites[168]
- Neptune's Moons[169] by NASA's Solar System Exploration[170]
- Gazetteer of Planetary Nomenclature—Neptune (USGS)[171]
- Simulation showing the position of Neptune's Moon[172]
- The 13 Moons Of Neptune – Astronoo[173]

Planetary rings

Rings of Neptune

The **rings of Neptune** consist primarily of five principal rings and were first discovered (as "arcs") on 22 July 1984 in Chile by Patrice Bouchet, Reinhold Häfner and Jean Manfroid at La Silla Observatory (ESO) during an observing program proposed by André Brahic and Bruno Sicardy from Paris Observatory, and at Cerro Tololo Interamerican Observatory by F. Vilas and L.-R. Elicer for a program led by William Hubbard. They were eventually imaged in 1989 by the *Voyager 2* spacecraft. At their densest, they are comparable to the less dense portions of Saturn's main rings such as the C ring and the Cassini Division, but much of Neptune's ring system is quite tenuous, faint and dusty, more closely resembling the rings of Jupiter. Neptune's rings are named after astronomers who contributed important work on the planet: Galle, Le Verrier, Lassell, Arago, and Adams.[174] Neptune also has a faint unnamed ring coincident with the orbit of the moon Galatea. Three other moons orbit between the rings: Naiad, Thalassa and Despina.

The rings of Neptune are made of extremely dark material, likely organic compounds processed by radiation, similar to that found in the rings of Uranus. The proportion of dust in the rings (between 20% and 70%) is high, while their optical depth is low to moderate, at less than 0.1. Uniquely, the Adams ring includes five distinct arcs, named Fraternité, Égalité 1 and 2, Liberté, and Courage. The arcs occupy a narrow range of orbital longitudes and are remarkably stable, having changed only slightly since their initial detection in 1980. How the arcs are stabilized is still under debate. However, their stability is probably related to the resonant interaction between the Adams ring and its inner shepherd moon, Galatea.

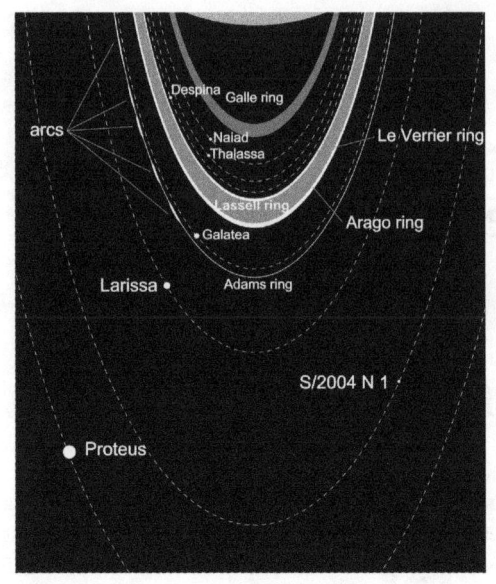

Figure 70: *The scheme of Neptune's ring-moon system. Solid lines denote rings; dashed lines denote orbits of moons.*

Discovery and observations

The first mention of rings around Neptune dates back to 1846 when William Lassell, the discoverer of Neptune's largest moon, Triton, thought he had seen a ring around the planet. However, his claim was never confirmed and it is likely that it was an observational artifact. The first reliable detection of a ring was made in 1968 by stellar occultation, although that result would go unnoticed until 1977 when the rings of Uranus were discovered. Soon after the Uranus discovery, a team from Villanova University led by Harold J. Reitsema began searching for rings around Neptune. On 24 May 1981, they detected a dip in a star's brightness during one occultation; however, the manner in which the star dimmed did not suggest a ring. Later, after the Voyager fly-by, it was found that the occultation was due to the small Neptunian moon Larissa, a highly unusual event.

In the 1980s, significant occultations were much rarer for Neptune than for Uranus, which lay near the Milky Way at the time and was thus moving against a denser field of stars. Neptune's next occultation, on 12 September 1983, resulted in a possible detection of a ring. However, ground-based results were inconclusive. Over the next six years, approximately 50 other occultations were observed with only about one-third of them yielding positive results.

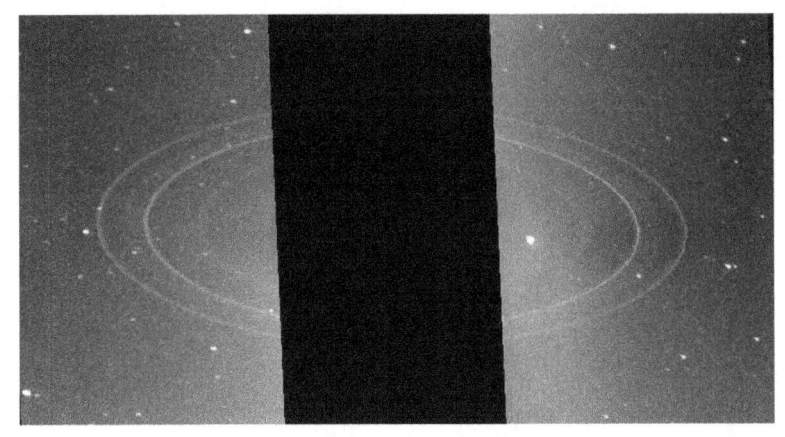

Figure 71: *A pair of Voyager 2 images of Neptune's ring system*

Something (probably incomplete arcs) definitely existed around Neptune, but the features of the ring system remained a mystery. The *Voyager 2* spacecraft made the definitive discovery of the Neptunian rings during its fly-by of Neptune in 1989, passing by as close as 4,950 km (3,080 mi) above the planet's atmosphere on 25 August. It confirmed that occasional occultation events observed before were indeed caused by the arcs within the Adams ring (see below). After the *Voyager* fly-by the previous terrestrial occultation observations were reanalyzed yielding features of the ring's arcs as they were in 1980s, which matched those found by *Voyager 2* almost perfectly.

Since *Voyager 2*'s fly-by, the brightest rings (Adams and Le Verrier) have been imaged with the Hubble Space Telescope and Earth-based telescopes, owing to advances in resolution and light-gathering power. They are visible, slightly above background noise levels, at methane-absorbed wavelengths in which the glare from Neptune is significantly reduced. The fainter rings are still far below the visibility threshold.

General properties

Neptune possesses five distinct rings named, in order of increasing distance from the planet, Galle, Le Verrier, Lassell, Arago and Adams. In addition to these well-defined rings, Neptune may also possess an extremely faint sheet of material stretching inward from the Le Verrier to the Galle ring, and possibly farther in toward the planet. Three of the Neptunian rings are narrow, with widths of about 100 km or less; in contrast, the Galle and Lassell rings are broad—their widths are between 2,000 and 5,000 km. The Adams ring

Figure 72: *A Voyager ring image shown at increased brightness to bring out fainter features*

consists of five bright arcs embedded in a fainter continuous ring. Proceeding counterclockwise, the arcs are: Fraternité, Égalité 1 and 2, Liberté, and Courage. The first three names come from "liberty, equality, fraternity", the motto of the French Revolution and Republic. The terminology was suggested by their original discoverers, who had found them during stellar occultations in 1984 and 1985. Four small Neptunian moons have orbits inside the ring system: Naiad and Thalassa orbit in the gap between the Galle and Le Verrier rings; Despina is just inward of the Le Verrier ring; and Galatea lies slightly inward of the Adams ring, embedded in an unnamed faint, narrow ringlet.

The Neptunian rings contain a large quantity of micrometer-sized dust: the dust fraction by cross-section area is between 20% and 70%. In this respect they are similar to the rings of Jupiter, in which the dust fraction is 50%–100%, and are very different from the rings of Saturn and Uranus, which contain little dust (less than 0.1%). The particles in Neptune's rings are made from a dark material; probably a mixture of ice with radiation-processed organics. The rings are reddish in color, and their geometrical (0.05) and Bond (0.01–0.02) albedos are similar to those of the Uranian rings' particles and the inner Neptunian moons. The rings are generally optically thin (transparent); their normal optical depths do not exceed 0.1. As a whole, the Neptunian rings resemble those of Jupiter; both systems consist of faint, narrow, dusty ringlets and even fainter broad dusty rings.

The rings of Neptune, like those of Uranus, are thought to be relatively young; their age is probably significantly less than that of the Solar System. Also, like those of Uranus, Neptune's rings probably resulted from the collisional fragmentation of onetime inner moons. Such events create moonlet belts, which act as the sources of dust for the rings. In this respect the rings of Neptune are similar to faint dusty bands observed by *Voyager 2* between the main rings of Uranus.

Inner rings

The innermost ring of Neptune is called the *Galle ring* after Johann Gottfried Galle, the first person to see Neptune through a telescope (1846). It is about 2,000 km wide and orbits 41,000–43,000 km from the planet. It is a faint ring with an average normal optical depth of around 10^{-4},[175] and with an equivalent depth of 0.15 km.[176] The fraction of dust in this ring is estimated from 40% to 70%.

The next ring is named the *Le Verrier ring* after Urbain Le Verrier, who predicted Neptune's position in 1846. With an orbital radius of about 53,200 km, it is narrow, with a width of about 113 km. Its normal optical depth is 0.0062 ± 0.0015, which corresponds to an equivalent depth of 0.7 ± 0.2 km. The dust fraction in the Le Verrier ring ranges from 40% to 70%. The small moon Despina, which orbits just inside of it at 52,526 km, may play a role in the ring's confinement by acting as a shepherd.

The *Lassell ring*, also known as the *plateau*, is the broadest ring in the Neptunian system. It is the namesake of William Lassell, the English astronomer who discovered Neptune's largest moon, Triton. This ring is a faint sheet of material occupying the space between the Le Verrier ring at about 53,200 km and the Arago ring at 57,200 km. Its average normal optical depth is around 10^{-4}, which corresponds to an equivalent depth of 0.4 km. The ring's dust fraction is in the range from 20% to 40%.

There is a small peak of brightness near the outer edge of the Lassell ring, located at 57,200 km from Neptune and less than 100 km wide, which some planetary scientists call the *Arago ring* after François Arago, a French mathematician, physicist, astronomer and politician. However, many publications do not mention the Arago ring at all.

Figure 73: *Arcs in the Adams ring (left to right: Fraternité, Égalité, Liberté), plus the Le Verrier ring on the inside*

Adams ring

The outer Adams ring, with an orbital radius of about 63,930 km, is the best studied of Neptune's rings. It is named after John Couch Adams, who predicted the position of Neptune independently of Le Verrier. This ring is narrow, slightly eccentric and inclined, with total width of about 35 km (15–50 km), and its normal optical depth is around 0.011 ± 0.003 outside the arcs, which corresponds to the equivalent depth of about 0.4 km. The fraction of dust in this ring is from 20% to 40%—lower than in other narrow rings. Neptune's small moon Galatea, which orbits just inside of the Adams ring at 61,953 km, acts like a shepherd, keeping ring particles inside a narrow range of orbital radii through a 42:43 outer Lindblad resonance. Galatea's gravitational influence creates 42 radial wiggles in the Adams ring with an amplitude of about 30 km, which have been used to infer Galatea's mass.

Arcs

The brightest parts of the Adams ring, the ring arcs, were the first elements of Neptune's ring system to be discovered. The arcs are discrete regions within the ring in which the particles that it comprises are mysteriously clustered together. The Adams ring is known to comprise five short arcs, which occupy a relatively narrow range of longitudes from 247° to 294°.[177] In 1986 they were located between longitudes of:

- 247–257° (Fraternité),
- 261–264° (Égalité 1),
- 265–266° (Égalité 2),
- 276–280° (Liberté),
- 284.5–285.5° (Courage).

The brightest and longest arc was Fraternité; the faintest was Courage. The normal optical depths of the arcs are estimated to lie in the range 0.03–0.09 (0.034 ± 0.005 for the leading edge of Liberté arc as measured by stellar occultation); the radial widths are approximately the same as those of the continuous ring—about 30 km. The equivalent depths of arcs vary in the range 1.25–2.15 km (0.77 ± 0.13 km for the leading edge of Liberté arc). The fraction of dust in the arcs is from 40% to 70%. The arcs in the Adams ring are somewhat similar to the arc in Saturn's G ring.

The highest resolution *Voyager 2* images revealed a pronounced clumpiness in the arcs, with a typical separation between visible clumps of 0.1° to 0.2°, which corresponds to 100–200 km along the ring. Because the clumps were not resolved, they may or may not include larger bodies, but are certainly associated with concentrations of microscopic dust as evidenced by their enhanced brightness when backlit by the Sun.

The arcs are quite stable structures. They were detected by ground-based stellar occultations in the 1980s, by *Voyager 2* in 1989 and by Hubble Space Telescope and ground-based telescopes in 1997–2005 and remained at approximately the same orbital longitudes. However some changes have been noticed. The overall brightness of arcs decreased since 1986. The Courage arc jumped forward by 8° to 294° (it probably jumped over to the next stable co-rotation resonance position) while the Liberté arc had almost disappeared by 2003. The Fraternité and Égalité (1 and 2) arcs have demonstrated irregular variations in their relative brightness. Their observed dynamics is probably related to the exchange of dust between them. Courage, a very faint arc found during the Voyager flyby, was seen to flare in brightness in 1998; it was back to its usual dimness by June 2005. Visible light observations show that the total amount of material in the arcs has remained approximately constant, but they are dimmer in the infrared light wavelengths where previous observations were taken.

Confinement

The arcs in the Adams ring remain unexplained. Their existence is a puzzle because basic orbital dynamics imply that they should spread out into a uniform ring over a matter of years. Several theories about the arcs' confinement have been suggested, the most widely publicized of which holds that Galatea confines the arcs via its 42:43 co-rotational inclination resonance (CIR).[178] The resonance creates 84 stable sites along the ring's orbit, each 4° long, with arcs residing in the adjacent sites. However measurements of the rings' mean motion with Hubble and Keck telescopes in 1998 led to the conclusion that the rings are not in CIR with Galatea.

A later model suggested that confinement resulted from a co-rotational eccentricity resonance (CER).[179] The model takes into account the finite mass of the Adams ring, which is necessary to move the resonance closer to the ring. A byproduct of this theory is a mass estimate for the Adams ring—about 0.002 of the mass of Galatea. A third theory proposed in 1986 requires an additional moon orbiting inside the ring; the arcs in this case are trapped in its stable Lagrangian points. However *Voyager 2*'s observations placed strict constraints on the size and mass of any undiscovered moons, making such a theory unlikely. Some other more complicated theories hold that a number of moonlets are trapped in co-rotational resonances with Galatea, providing confinement of the arcs and simultaneously serving as sources of the dust.

Exploration

The rings were investigated in detail during the *Voyager 2* spacecraft's flyby of Neptune in August 1989. They were studied with optical imaging, and through observations of occultations in ultraviolet and visible light. The spaceprobe observed the rings in different geometries relative to the Sun, producing images of back-scattered, forward-scattered and side-scattered light.[180] Analysis of these images allowed derivation of the phase function (dependence of the ring's reflectivity on the angle between the observer and Sun), and geometrical and Bond albedo of ring particles. Analysis of Voyager's images also led to discovery of six inner moons of Neptune, including the Adams ring shepherd Galatea.

Properties

Ring name	Radius (km)	Width (km)	Eq. depth (km)[181]	N. Opt. depth	Dust fraction,%	Ecc.	Incl.(°)	Notes
Galle (N42)	40,900–42,900	2,000	0.15	$\sim 10^{-4}$	40–70	?	?	Broad faint ring
Le Verrier (N53)	53,200 ± 20	113	0.7 ± 0.2	$6.2 \pm 1.5 \times 10^{-3}$	40–70	?	?	Narrow ring
Lassell	53,200–57,200	4,000	0.4	$\sim 10^{-4}$	20–40	?	?	Lassell ring is a faint sheet of material stretching from Le Verrier to Arago
Arago	57,200	<100	?	?	?	?	?	
Adams (N63)	62,932 ± 2	15–50	0.4 1.25–2.15 (in arcs)	0.011 ± 0.003 0.03–0.09 (in arcs)	20–40 40–70 (in arcs)	$4.7 \pm 0.2 \times 10^{-4}$	0.0617 ± 0.0043	Five bright arcs

*A question mark means that the parameter is not known.

References
External links

> Wikimedia Commons has media related to *Rings of Neptune*.

- Neptune's Rings[182] by NASA's Solar System Exploration[183]
- Gazetteer of Planetary Nomenclature – Ring and Ring Gap Nomenclature (Neptune), USGS[184]

<indicator name="featured-star"> ⭐ </indicator>

Exploration

Exploration of Neptune

The **exploration of Neptune** has only begun with one spacecraft, *Voyager 2* in 1989. Currently there are no approved future missions to visit the Neptunian system. NASA, ESA and also independent academic groups have proposed future scientific missions to visit Neptune. Some mission plans are still active, while others have been abandoned or put on hold.

Neptune has also been scientifically studied from afar with telescopes, primarily since the mid 1990s. This includes the Hubble Space Telescope but most importantly ground-based telescopes using adaptive optics.

Voyager 2

After having visited Saturn successfully, it was decided to continue and fund further missions by *Voyager 2* to Uranus and Neptune. These missions were conducted by the Jet Propulsion Laboratory and the Neptunian mission was dubbed "Voyager Neptune Interstellar Mission". *Voyager 2*'s observation phase of Neptune began 5 June 1989, the spacecraft officially reached the Neptunian system on 25 August and the data collection stopped on 2 October.

On 25 August, in *Voyager 2*'s last planetary encounter, the spacecraft swooped only 4,950 km (3,080 mi) above Neptune's north pole, the closest approach it made to any body since it left Earth in 1977. When the spacecraft visited the Neptunian system, Neptune was the farthest known body in the Solar system. It was not until 1999, that Pluto was further from the Sun in its trajectory. *Voyager 2* studied Neptune's atmosphere, Neptune's rings, its magnetosphere, and Neptune's moons.[185] The Neptunian system had been studied scientifically for many years with telescopes and indirect methods, but the close inspection by

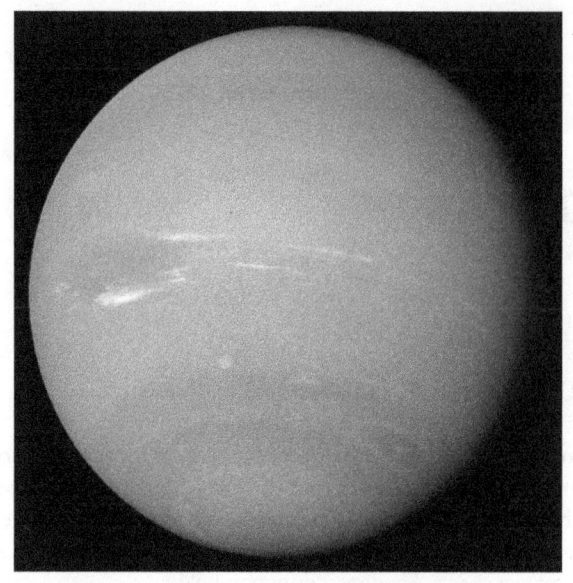

Figure 74: *Neptune. Processed image from Voyager 2's narrow-angle camera 16 or 17 of August 1989. Neptune's south pole is at the bottom of the image.*

Figure 75: *Voyager 2 image of Triton*

Exploration of Neptune

Figure 76: *Voyager 2 spacecraft*

the *Voyager 2* probe settled many issues and revealed a plethora of information that could not have been obtained otherwise. The data from *Voyager 2* are still the best data available on this planet in most cases.

The exploration mission revealed that Neptune's atmosphere is very dynamic, even though it receives only 3% of the sunlight that Jupiter receives. Winds on Neptune were found to be the strongest in the Solar System, up to three times stronger than Jupiter's and nine times stronger than the strongest winds on Earth. Most winds blew westward, opposite the planets rotation. Separate cloud decks were discovered, with cloud systems emerging and dissolving within hours and giant storms circling the entire planet in 16–18 hours in the upper layers. *Voyager 2* discovered an anticyclone dubbed the Great Dark Spot, similar to Jupiter's Great Red Spot. However, images taken by the Hubble Space Telescope in 1994, revealed that the Great Dark Spot had disappeared. Also seen in Neptune's upper atmosphere at that time was an almond-shaped spot designated D2, and a bright, quickly moving cloud high above the cloud decks, dubbed "Scooter".[186]

The fly-by of the Neptunian system provided the first accurate measurement of Neptune's mass which was found to be 0.5 percent less than previously calculated. The new figure disproved the hypothesis that an undiscovered Planet X acted upon the orbits of Neptune and Uranus.[187]

Neptune's magnetosphere was also studied by *Voyager 2*. The magnetic field was found to be highly tilted and largely offset from the planet's centre. The probe discovered auroras, but much weaker than those on Earth or other planets. The radio instruments on board found that Neptune's day lasts 16 hours

Figure 77: *Voyager 2 image of Proteus*

and 6.7 minutes. Neptune's rings had been observed from Earth many years prior to *Voyager 2*'s visit, but the close inspection revealed that the ring systems were full circle and intact, and a total of four rings were counted.

Voyager 2 discovered six new small moons orbiting Neptune's equatorial plane, and they were named Naiad, Thalassa, Despina, Galatea, Larissa and Proteus. Only three of Neptune's moons were photographed in detail: Proteus, Nereid, and Triton; of which the last two were the only Neptunian moons known prior to the visit. Proteus turned out to be an ellipsoid, as large as gravity allows an ellipsoid body to become without rounding into a sphere. It appeared very dark in color, almost like soot. Nereid, though discovered in 1949, still has very little known about it. Triton was revealed as having a remarkably active past. Active geysers and polar caps were discovered and a very thin atmosphere as well, with thin clouds of what is thought to be nitrogen ice particles. At just 38 K (–235.2 °C), it is the coldest known planetary body in the Solar System. The closest approach to Triton was about 40,000 km (25,000 mi) and it was the last solid world that *Voyager 2* explored close-by.

Summary of missions to the outer Solar System

System Spacecraft	Jupiter	Saturn	Uranus	Neptune	Pluto
Pioneer 10	**1973** flyby				
Pioneer 11	**1974** flyby	**1979** flyby			
Voyager 1	**1979** flyby	**1980** flyby			
Voyager 2	**1979** flyby	**1981** flyby	**1986** flyby	**1989** flyby	
Galileo	**1995–2003** orbiter; **1995, 2003** atmospheric				
Ulysses	**1992, 2004** gravity assist				
Cassini–Huygens	**2000** gravity assist	**2004–2017** orbiter; **2005** Titan lander			
New Horizons	**2007** gravity assist				**2015** flyby
Juno	**2016–** orbiter				
Jupiter Icy Moons Explorer	**2022–** Planned orbiter				
Europa Clipper	**2025–** Planned orbiter				

Future missions

Currently there are no approved future missions to visit the Neptunian system. NASA, ESA and independent academic groups have proposed and worked on future scientific missions to visit Neptune. Some mission plans are still active, while others have been abandoned or put on hold.

After the *Voyager* flyby, NASA's next step in scientific exploration of the Neptune system is considered to be a flagship orbiter mission. Such a hypothetical mission is envisioned to be possible in the late 2020s or early 2030s. Another one proposed for the 2040s is called the Neptune-Triton Explorer (NTE). NASA has researched several other project options for both flyby and orbiter missions (of similar design as the *Cassini–Huygens* mission to Saturn). These missions are often collectively called "RMA Neptune-Triton-KBO" missions, which also includes orbital missions that would not visit Kuiper belt objects (KBOs). Because of budgetary constraints, technological considerations, scientific priorities and other factors, none of these have been approved.

Specific exploration missions to Neptune includes:

- Neptune Orbiter mission — An orbiting mission concept focusing on Neptune and Triton.
- *Argo* — A mission concept New Frontiers flyby mission to visit Jupiter, Saturn, Neptune (with Triton) and the Kuiper belt with launch in 2019.
- ODINUS — A mission concept based on a twin spacecraft mission to investigate the Neptunian and Uranian systems. Launch date would be 2034.
- OSS mission — A proposed collaborative flyby mission by ESA and NASA. Its main focus would be to map the gravitational fields in deep space, including the Outer Solar System (up to 50 AU).

Direct flyby missions to Neptune are only preferable in window gaps with a 12-year interval, due to the complexity of planetary positions. There is currently a window open for launching a Neptune mission from 2014 to 2019, with the next opportunity occurring from 2031. These constraints are based on current rocket technology which relies on gravity assists from Jupiter and Saturn. With the new Space Launch System (SLS) technology in development at Boeing, deep space missions with heavier payloads could potentially be propelled at much greater speeds (200 AU in 15 years) and missions to the outer planets could be launched independently of gravity assistance.

Scientific studies from afar

Space telescopes such as the Hubble Space Telescope has signified a new era of detailed observations of faint objects from afar in all of the electromagnetic spectrum. This includes faint objects in the Solar system, such as Neptune. Adaptive optics technology has also allowed for detailed observations of faint objects from ground-based telescopes and this technology has been used since 1997 to obtain scientific information on Neptune and its atmosphere. These image recordings now exceed the capability of HST by far and in some instances even the Voyager images, such as Uranus.[188] Ground-based observations are however always limited in their registration of electromagnetic waves of certain wavelengths, due to the inevitable atmospheric absorption, in particular of high energy waves.[189]

Sources

- Neptune[190] *Voyager 2* - The Interstellar Mission, Jet Propulsion Laboratory, California Institute of Technology
- Neptune: In Depth[191] Planets, NASA

External links

- 25 Years After Neptune: Reflections on Voyager[192] NASA Voyager website
- Images of Neptune and All Available Satellites[193] Photojournal, JPL

Appendix

References

[1] —Select "Ephemeris Type: Orbital Elements", "Time Span: 2000-01-01 12:00 to 2000-01-02". ("Target Body: Neptune Barycenter" and "Center: Solar System Barycenter (@0)".)

[2] Orbital elements refer to the Neptune barycentre and Solar System barycentre. These are the instantaneous osculating values at the precise J2000 epoch. Barycentre quantities are given because, in contrast to the planetary centre, they do not experience appreciable changes on a day-to-day basis from the motion of the moons.

[3] (produced with Solex 10 http://chemistry.unina.it/~alvitagl/solex/ written by Aldo Vitagliano; see also Invariable plane)

[4]

[5] Based on the volume within the level of 1 bar atmospheric pressure

[6] Neptune is denser and physically smaller than Uranus because Neptune's greater mass gravitationally compresses the atmosphere more.

[7] Moore (2000):206

[8] See also the Greek article about the planet.

[9] "Appendix 5: Planetary Linguistics" http://nineplanets.org/days.html, Nineplanets.org

[10] The mass of Earth is 5.9736 kg, giving a mass ratio
UNIQ-math-0-0235c40ef49a14df-QINU
The mass of Uranus is 8.6810 kg, giving a mass ratio
UNIQ-math-1-0235c40ef49a14df-QINU
The mass of Jupiter is 1.8986 kg, giving a mass ratio
UNIQ-math-2-0235c40ef49a14df-QINU
Mass values from

[11] Burgess (1991):64–70.

[12] Imke de Pater and Jack J. Lissauer (2001), *Planetary Sciences* https//books.google.com, 1st edition, page 224.

[13] Jean Meeus, *Astronomical Algorithms* (Richmond, VA: Willmann-Bell, 1998) 273. Supplemented by further use of VSOP87. The last three aphelia were 30.33 AU, the next is 30.34 AU. The perihelia are even more stable at 29.81 AU

[14] (Bill Folkner at JPL) https://twitter.com/elakdawalla/status/21525820626

[15] —Numbers generated using the Solar System Dynamics Group, Horizons On-Line Ephemeris System.

[16] Hubble Space Telescope discovers fourteenth tiny moon orbiting Neptune l Space, Military and Medicine http://www.news.com.au/technology/sci-tech/hubble-space-telescope-discovers-fourteenth-tiny-moon-orbiting-neptune/story-fn5fsgyc-1226679913807. News.com.au (16 July 2013). Retrieved on 28 July 2013.

[17] Mass of Triton: 2.14 kg. Combined mass of 12 other known moons of Neptune: 7.53 kg, or 0.35%. The mass of the rings is negligible.

[18] UNIQ-math-3-0235c40ef49a14df-QINU

[19] See the respective articles for magnitude data.

[20] Moore (2000):207.

[21] In 1977, for example, even the rotation period of Neptune remained uncertain.

[22] First Ground-Based Adaptive Optics Observations of Neptune and Proteus http://citeseerx.ist.psu.edu/viewdoc/download?doi=10.1.1.66.7754&rep=rep1&type=pdf Planetary & Space Science Vol. 45, No. 8, pp. 1031-1036, 1997

[23] Uranus and Neptune https://books.google.com/books?id=8s1JV-TrXacC&pg=PA147#v=onepage&q=Neptune&f=false Reports on Astronomy 2003-2005, p. 147 f.

[24] Burgess (1991):46–55.

[25] Tom Standage (2000). *The Neptune File: A Story of Astronomical Rivalry and the Pioneers of Planet Hunting.* New York: Walker. p. 188.

[26] http://nssdc.gsfc.nasa.gov/planetary/factsheet/neptunefact.html

[27] http://www.nineplanets.org/neptune.html
[28] http://www.astronomycast.com/astronomy/episode-63-neptune/
[29] http://solarsystem.nasa.gov/planets/profile.cfm?Object=Neptune
[30] http://solarsystem.nasa.gov
[31] http://www.projectshum.org/Planets/neptune.html
[32] http://www.sixtysymbols.com/videos/neptune.htm
[33] http://www.planetary.org/blogs/guest-blogs/2013/neptune-the-new-amateur-boundary.html
[34] Bouvard (1821)
[35] [Anon.] (2001) "Bouvard, Alexis", *Encyclopædia Britannica*, Deluxe CDROM edition
[36] Sampson (1904)
[37] Dennis Rawlins, *Bulletin of the American Astronomical Society*, volume 16, page 734, 1984 (first publication of British astronomer J.Hind's charge that Adams's secrecy disallows his claim).
[38] Robert Smith, *Isis*, volume 80, pages 395–422, September, 1989
[39] Smart (1947) *p.59*
[40] Adams's final prediction on 2 September 1846 was for a true longitude of about 315 degrees. That was 12 degrees west of Neptune. The large error was first emphasized in Adams's exact calculation of his prediction of 315 degrees was recovered in 2010 http://www.dioi.org/cot.htm#hpfp.
[41] "A brief History of Astronomy in Berlin and the Wilhelm-Foerster-Observatory" (accessed September 23rd 2010) http://www.planetarium-berlin.de/pages/hist/WFS-History.html
[42] Astronomy in Berlin: Johann Friedrich Galle (accessed September 25th 2010) http://bdaugherty.tripod.com/astronomy/berlin.html#GALLE
[43] Frommers: Deutsches Museum (accessed September 25th 2010) http://www.frommers.com/micro/2010/35-places-to-take-your-kids-before-they-grow-up/deutches-museum.html
[44] http://adsabs.harvard.edu/full/1992JHA....23..261H
[45] https://eee.uci.edu/clients/bjbecker/ExploringtheCosmos/lecture12.html
[46] http://adsabs.harvard.edu/abs/1847MmRAS..16..385A
[47] http://www.gutenberg.net/etext/10655
[48] http://adsabs.harvard.edu/cgi-bin/nph-bib_query?bibcode=1821tapp.book.....B&db_key=AST&data_type=HTML&format=&high=44b52c369020669
[49] http://adsabs.harvard.edu/abs/1988JHA....19..121C
[50] //doi.org/10.1098/rsnr.1996.0027
[51] http://www.oxforddnb.com/view/article/123
[52] http://www.oxforddnb.com/help/subscribe#public
[53] http://adsabs.harvard.edu/abs/1883MNRAS..43..160.
[54] //doi.org/10.1093/mnras/43.4.160
[55] https://web.archive.org/web/20051111190351/http://www.ucl.ac.uk/sts/nk/neptune/index.htm
[56] http://www.ucl.ac.uk/sts/nk/neptune/index.htm
[57] http://www-history.mcs.st-andrews.ac.uk/history/HistTopics/Neptune_and_Pluto.html
[58] http://www.dioi.org/vols/w23.pdf
[59] http://www.dioi.org/vols/w42.pdf
[60] http://adsabs.harvard.edu/abs/1994DIO.....4...92R
[61] http://www.dioi.org/vols/w91.pdf
[62] http://adsabs.harvard.edu/abs/1904MmRAS..54..143S
[63] //doi.org/10.1098/rsnr.2007.0187
[64] http://www.sciam.com/article.cfm?articleID=000CA850-8EA4-119B-8EA483414B7FFE9F
[65] http://adsabs.harvard.edu/abs/1946Natur.158..648S
[66] //doi.org/10.1038/158648a0
[67] //doi.org/10.1086/355082
[68] //doi.org/10.2307/3965133
[69] //www.jstor.org/stable/3965133
[70] http://www.cfeps.net/
[71] //en.wikipedia.org/w/index.php?title=Template:Distant_minor_planets_sidebar&action=edit
[72] Kuiper belt – oxforddictionaries.com http://www.oxforddictionaries.com/definition/english/Kuiper-belt

[73] Johnson, Torrence V.; and Lunine, Jonathan I.; *Saturn's moon Phoebe as a captured body from the outer Solar System*, Nature, Vol. 435, pp. 69–71

[74] NEW HORIZONS *The PI's Perspective* http://pluto.jhuapl.edu/overview/piPerspective.php?page=piPerspective_08_24_2012

[75] The literature is inconsistent in the usage of the terms *scattered disc* and *Kuiper belt*. For some, they are distinct populations; for others, the scattered disc is part of the Kuiper belt. Authors may even switch between these two uses in one publication.<ref>Weissman and Johnson, 2007, *Encyclopedia of the solar system*, footnote p. 584

[76] Randall 2015, p. 106.

[77] Davies, p. 2

[78] Davies, p. 14

[79] Davies p. 38

[80] Randall 2015, p. 105.

[81] Davies p. 39

[82] Davies p. 191

[83] Davies p. 50

[84] Davies p. 51

[85] Davies pp. 52, 54, 56

[86] Davies pp. 57, 62

[87] Davies p. 65

[88] Davies p. 199

[89] Clyde Tombaugh, "The Last Word", Letters to the Editor, *Sky & Telescope*, December 1994, p. 8

[90] Davies p. 104

[91] Davies p. 107

[92] Davies p. 118

[93] New Horizons' catches a wandering Kuiper Belt Object not far off http://www.spacedaily.com/reports/New_Horizons_catches_a_wandering_Kuiper_Belt_Object_not_far_off_999.html spacedaily.com Laurel MD (SPX). December 7, 2015.

[94] http://www2.ess.ucla.edu/~jewitt/kb.html

[95] http://www2.ess.ucla.edu/~jewitt/kb/gerard.html

[96] http://www.physics.ucf.edu/~yfernandez/cometlist.html

[97] http://solarsystem.nasa.gov/planets/profile.cfm?Object=KBOs

[98] http://solarsystem.nasa.gov

[99] http://www.boulder.swri.edu/ekonews/

[100] http://www.johnstonsarchive.net/astro/tnos.html

[101] http://www.minorplanetcenter.org/iau/lists/OuterPlot.html

[102] http://www.iau.org/

[103] https://web.archive.org/web/20030410073742/http://www.astronomy2006.com/

[104] http://www.nature.com/nature/journal/v424/n6949/fig_tab/nature01725_F1.html

[105] //doi.org/10.1038/nature01725

[106] //www.ncbi.nlm.nih.gov/pubmed/12904784

[107] http://www.space.com/scienceastronomy/060814_tno_found.html

[108] http://www.astronomycast.com/astronomy/episode-64-pluto-and-the-icy-outer-solar-system/

[109] https://web.archive.org/web/20160304083511/http://canadapodcasts.ca/podcasts/TheDays/1358804?start=0

[110] http://nineplanets.org/kboc.html

[111] http://www.minorplanetcenter.net/iau/lists/TNOs.html

[112] //en.wikipedia.org/w/index.php?title=Template:Distant_minor_planets_sidebar&action=edit

[113] E. I. Chiang and Y. Lithwick *Neptune Trojans as a Testbed for Planet Formation*, The Astrophysical Journal, **628**, pp. 520–532 Preprint http://www.arxiv.org/abs/astro-ph/0502276

[114] After the asteroid belt, the Jupiter trojans, the trans-Neptunian objects and the Mars trojans.

[115]

[116] https://minorplanetcenter.net/db_search/show_object?object_id=2001+QR322

[117] https://minorplanetcenter.net/db_search/show_object?object_id=385571

[118] https://minorplanetcenter.net/db_search/show_object?object_id=2005+TN53

[119] https://minorplanetcenter.net/db_search/show_object?object_id=385695
[120] https://minorplanetcenter.net/db_search/show_object?object_id=2006+RJ103
[121] https://minorplanetcenter.net/db_search/show_object?object_id=2007+VL305
[122] https://minorplanetcenter.net/db_search/show_object?object_id=2008+LC18
[123] https://minorplanetcenter.net/db_search/show_object?object_id=2004+KV18
[124] https://minorplanetcenter.net/db_search/show_object?object_id=316179
[125] https://minorplanetcenter.net/db_search/show_object?object_id=2010+TS191
[126] https://minorplanetcenter.net/db_search/show_object?object_id=2010+TT191
[127] https://minorplanetcenter.net/db_search/show_object?object_id=2011+HM102
[128] https://minorplanetcenter.net/db_search/show_object?object_id=2011+SO277
[129] https://minorplanetcenter.net/db_search/show_object?object_id=2011+WG157
[130] https://minorplanetcenter.net/db_search/show_object?object_id=2012+UV177
[131] https://minorplanetcenter.net/db_search/show_object?object_id=2013+KY18
[132] https://minorplanetcenter.net/db_search/show_object?object_id=2014+QO441
[133] https://minorplanetcenter.net/db_search/show_object?object_id=2014+QP441
[134] MPEC 2005-U97 : 2005 TN74, 2005 TO74 http://www.minorplanetcenter.net/iau/mpec/K05/K05U97.html Minor Planet Center
[135] //arxiv.org/abs/1007.2541
[136] http://adsabs.harvard.edu/abs/2010IJAsB...9..227H
[137] //doi.org/10.1017/S1473550410000212
[138] https://www.newscientist.com/article/dn9340-new-trojan-asteroid-hints-at-huge-neptunian-cloud.html
[139] An astronomical unit, or AU, is the average distance between the Earth and the Sun, or about 150 million kilometres. It is the standard unit of measurement for interplanetary distances.
[140] Zeilik & Gregory 1998, p. 207.
[141] The combined mass of Jupiter, Saturn, Uranus and Neptune is 445.6 Earth masses. The mass of remaining material is ∼5.26 Earth masses or 1.1% (see Solar System#Notes and List of Solar System objects by mass)
[142] The reason that Saturn, Uranus and Neptune all moved outward whereas Jupiter moved inward is that Jupiter is massive enough to eject planetesimals from the Solar System, while the other three outer planets are not. To eject an object from the Solar System, Jupiter transfers energy to it, and so loses some of its own orbital energy and moves inwards. When Neptune, Uranus and Saturn perturb planetesimals outwards, those planetesimals end up in highly eccentric but still bound orbits, and so can return to the perturbing planet and possibly return its lost energy. On the other hand, when Neptune, Uranus and Saturn perturb objects inwards, those planets gain energy by doing so and therefore move outwards. More importantly, an object being perturbed inwards stands a greater chance of encountering Jupiter and being ejected from the Solar System, in which case the energy gains of Neptune, Uranus and Saturn obtained from their inwards deflections of the ejected object become permanent.
[143]

See also
[144] Zeilik & Gregory 1998, pp. 118–120.
[145] In all of these cases of transfer of angular momentum and energy, the angular momentum of the two-body system is conserved. In contrast, the summed energy of the moon's revolution plus the primary's rotation is not conserved, but decreases over time, due to dissipation via frictional heat generated by the movement of the tidal bulge through the body of the primary. If the primary were a frictionless ideal fluid, the tidal bulge would be centered under the satellite, and no transfer would take place. It is the loss of dynamical energy through friction that makes transfer of angular momentum possible.
[146] Duncan & Lissauer 1997.
[147] Zeilik & Gregory 1998, p. 320–321.
[148] Zeilik & Gregory 1998, p. 322.
[149] http://adsabs.harvard.edu/abs/1997Icar..125....1D
[150] //doi.org/10.1006/icar.1996.5568
[151] http://arquivo.pt/wayback/20160520023943/http://media.skyandtelescope.com/video/Solar_System_Sim.mov

[152] http://www.skyandtelescope.com
[153] http://www.cfa.harvard.edu/seuforum/animations/animations/galaxycollision.mpg
[154] http://www.space.com/common/media/video/player.php?videoRef=mm32_SunDeath
[155] Turrini & Marzari, 2008, *Phoebe and Saturn's irregular satellites: implications for the collisional capture scenario* http://www.saturnaftercassini.org/files/2_Turrini_Diego_A.pdf
[156] http://www.skyandtelescope.com/skytel/beyondthepage/8594717.html
[157] http://www.eurekalert.org/pub_releases/2007-12/asu-sss121107.php
[158] This is a IAU guideline that will be followed at the naming of every Neptunian moon, although one (S/2004 N 1) has yet to receive a permanent name.
[159] http://ssd.jpl.nasa.gov/?sat_elem#neptune (as of Dec-2014)
[160] The geometric albedo of an astronomical body is the ratio of its actual brightness at zero phase angle (i.e. as seen from the light source) to that of an idealized flat, fully reflecting, diffusively scattering (Lambertian) disk with the same cross-section. The Bond albedo, named after the American astronomer George Phillips Bond (1825–1865), who originally proposed it, is the fraction of power in the total electromagnetic radiation incident on an astronomical body that is scattered back out into space. The Bond albedo is a value strictly between 0 and 1, as it includes all possible scattered light (but not radiation from the body itself). This is in contrast to other definitions of albedo such as the geometric albedo, which can be above 1. In general, though, the Bond albedo may be greater or smaller than the geometric albedo, depending on surface and atmospheric properties of the body in question.
[161] Binary objects, objects with moons such as the Pluto–Charon system, are quite common among the larger trans-Neptunian objects (TNOs). Around 11% of all TNOs may be binaries. UNIQ-ref-0-0235c40ef49a14df-QINU
[162] Order refers to the position among other moons with respect to their average distance from Neptune.
[163] Label refers to the Roman numeral attributed to each moon in order of their discovery. This is a IAU guideline that will be followed at the naming of every Neptunian moon, although one (S/2004 N 1) has yet to receive a permanent name.
[164] Diameters with multiple entries such as "$60 \times 40 \times 34$" reflect that the body is not spherical and that each of its dimensions has been measured well enough to provide a 3-axis estimate. The dimensions of the five inner moons were taken from Karkoschka, 2003.http://ssd.jpl.nasa.gov/?sat_elem#neptune (as of Dec-2014) Dimensions of Proteus are from Stooke (1994). Dimensions of Triton are from Thomas, 2000, whereas its diameter is taken from Davies et al., 1991. The size of Nereid is from Smith, 1989. The sizes of the outer moons are from Sheppard et al., 2006.
[165] Mass of all moons of Neptune except Triton were calculated assuming a density of 1.3 g/cm^3. The volumes of Larissa and Proteus were taken from Stooke (1994). The mass of Triton is from Jacobson, 2009.
[166] Each moon's inclination is given relative to its local Laplace plane. Inclinations greater than 90° indicate retrograde orbits (in the direction opposite to the planet's rotation).
[167] Jones, Daniel (2003) [1917], Peter Roach, James Hartmann and Jane Setter, eds., *English Pronouncing Dictionary*, Cambridge: Cambridge University Press, ISBN 3-12-539683-2
[168] http://home.dtm.ciw.edu/users/sheppard/satellites/nepsatdata.html
[169] https://web.archive.org/web/20070609074953/http://solarsystem.nasa.gov/planets/profile.cfm?Object=Neptune&Display=Moons
[170] http://solarsystem.nasa.gov
[171] http://planetarynames.wr.usgs.gov/jsp/SystemSearch2.jsp?System=Neptune
[172] https://web.archive.org/web/20120303191038/http://www.orinetz.com/planet/tourprog/saturnmoons.html
[173] http://www.astronoo.com/articles/the13MoonsOfNeptune-en.html
[174] Listed in increasing distance from the planet
[175] The normal optical depth τ of a ring is the ratio of the total geometrical cross-section of the ring's particles to the area of the ring. It assumes values from zero to infinity. A light beam passing normally through a ring will be attenuated by the factor $e^{-\tau}$.
[176] The equivalent depth ED of a ring is defined as an integral of the normal optical depth across the ring. In other words UNIQ-nowiki-1-0235c40ef49a14df-QINU , where r is radius.

[177] The longitude system is fixed as of 18 August 1989. The zero point corresponds to the zero meridian on Neptune.

[178] The corotation inclination resonance (CIR) of the order m between a moon on inclined orbit and a ring occurs if the pattern speed of the perturbing potential UNIQ-math-4-0235c40ef49a14df-QINU (from a moon) equals the mean motion of the ring particles UNIQ-math-5-0235c40ef49a14df-QINU . In other words the following condition should be met UNIQ-math-6-0235c40ef49a14df-QINU , where UNIQ-math-7-0235c40ef49a14df-QINU and UNIQ-math-8-0235c40ef49a14df-QINU are the nodal precession rate and mean motion of the moon, respectively. CIR supports $2m$ stable sites along the ring.

[179] The corotation eccentricity resonance (CER) of the order m between a moon on eccentric orbit and a ring occurs if the pattern speed of the perturbing potential UNIQ-math-10-0235c40ef49a14df-QINU (from a moon) equals the mean motion of the ring particles UNIQ-math-11-0235c40ef49a14df-QINU . In other words the following condition should be met UNIQ-math-12-0235c40ef49a14df-QINU , where UNIQ-math-13-0235c40ef49a14df-QINU and UNIQ-math-14-0235c40ef49a14df-QINU are the apsidal precession rate and mean motion of the moon, respectively. CER supports m stable sites along the ring.

[180] Forward-scattered light is light scattered at a small angle relative to solar light. Back-scattered light is light scattered at an angle close to 180° (backwards) relative to solar light. The scattering angle is close to 90° for side-scattered light.

[181] The equivalent depth of Galle and Lassell rings is a product of their width and the normal optical depth.

[182] http://solarsystem.nasa.gov/planets/profile.cfm?Object=Neptune&Display=Rings

[183] http://solarsystem.nasa.gov

[184] http://planetarynames.wr.usgs.gov/append8.html

[185] See the "Neptune" page from JPL.

[186] See "Neptune:In Depth" from NASA.

[187] Tom Standage (2000). The Neptune File: A Story of Astronomical Rivalry and the Pioneers of Planet Hunting. New York: Walker. p. 188.

[188] Uranus and Neptune https://books.google.com/books?id=8s1JV-TrXacC&pg=PA147#v=onepage&q=Neptune&f=false Reports on Astronomy 2003-2005, p. 147 f.

[189] First Ground-Based Adaptive Optics Observations of Neptune and Proteus http://citeseerx.ist.psu.edu/viewdoc/download?doi=10.1.1.66.7754&rep=rep1&type=pdf *Planetary & Space Science* Vol. 45, No. 8, pp. 1031-1036, 1997

[190] http://voyager.jpl.nasa.gov/science/neptune.html

[191] http://solarsystem.nasa.gov/planets/neptune/indepth

[192] http://voyager.jpl.nasa.gov/news/25_years_after_neptune.html

[193] http://photojournal.jpl.nasa.gov/targetFamily/Neptune

Article Sources and Contributors

The sources listed for each article provide more detailed licensing information including the copyright status, the copyright owner, and the license conditions.

Neptune *Source:* https://en.wikipedia.org/w/index.php?oldid=856416778 *License:* Creative Commons Attribution-Share Alike 3.0 *Contributors:* A2soup, AaronMP84, Adelson Velsky Landis, Alanobrien, Ambi Valent, Arado, Archon 2488, Arjayay, Azmodes, BGManofID, Barjimoa, BatteryIncluded, Bender235, Bensational, Bill the Cat 7, BlueMoonlet, Bunkyray5, Cadiomals, Cancina5645, Coladar, CuriousEric, Cuttycuttiercuttiest, DN-boards1, DSmurf, DavideVeloria88, Dawnseeker2000, Dayirmiter, Deepanshu1707, Dinkytown, Dodshe, Dolphin33438, Double sharp, DrKay, Drbogdan, Edulovers, Ehrenkater, Et Cum Spiritu Tuo, Exoplanetaryscience, FerJox, Fotaun, Fountains of Bryn Mawr, Gallina3795, Gap9551, Georgelazenby, Glenn L, Go-in, GoPlayerJuggler, Gob Lofa, GrundyCamellia, Hbdragon88, Headbomb, Hillbillyholiday, HolyT, Huntster, Hz.tiang, Iggy the Swan, InedibleHulk, Inhighspeed, Isambard Kingdom, JPaestpreornJeolhlna, Jcpag2012, Jengelh, Jmencisom, Jmostly, John, Jon Kolbert, Jonesey95, JorisvS, Josve05a, Jrrs, JustAMuggle, KAP03, KajHansen, Kelvin13, Kheider, Koavf, Kwamikagami, LL221W, Labtek00, Lightlowemon, Lindenhurst Liberty, Lo Ximiendo, Maczkopeti, Martin Hogbin, MartinZ, Master of Time, Materialscientist, Mckburton, Me, Myself, and I are Here, Mezigue, Mikhail Ryazanov, Mscuthbert, Nedim Ardoğa, Neuralwarp, Nicolas Eynaud, Nihiltres, Nudelkopp17, Oldag07, Onore Baka Sama, Orange-kun, Orenburg1, Paintspot, Patteroast, Pdebee, PlayStation 14, PulauKakatua19, RPH, Randy Kryn, Reatlas, Redd Foxx 1991, Rfassbind, Rhatsa26X, RhinoMind, Richard75, Rivertorch, Rjwilmsi, Robert Skyhawk, RobertLovesPi, RockMagnetist (DCO visiting scholar), Roentgenium111, Rothorpe, Ryulong, Sailsbystars, SandyGeorgia, Saros136, Sbmeirow, SeL, SecretAssassin, Serendipodous, SeventeenTurtle, Sholokhov, Silenceisgod, SkoreKeep, Skynorth, Smk65536, SoylentPurple, Spiderjerky, Stamptrader, Steel1943, Stub Mandrel, Summstjr, TAnthony, Tetra quark, The Rambling Man, The Transhumanist, Thor Dockweiler, Tom.Reding, Tomruen, Trackteur, Trappist the monk, Twinsday, Tybomb124, Ufwuct, Vinci2005, Virtualphtn, Voello, W like wiki, Wasill37, Wcp07, WereSpielChequers, WolfmanSF, Writerwhiz2014, YoungDrac 1

Discovery of Neptune *Source:* https://en.wikipedia.org/w/index.php?oldid=853805435 *License:* Creative Commons Attribution-Share Alike 3.0 *Contributors:* A. Parrot, A8UDI, Ac44ck, Agent Smith (The Matrix), Aldebaran66, Allissonn, Antoni Barau, AstroLynx, AstroProf, Attilios, Bender235, Boccobrock, BornInLeningrad, Brandmeister (old), Bspeakmon, CapitalR, Chowbok, Chris55, Christian75, Citation bot 1, ClueBot NG, Cutler, Daryl Janzen, Dcirovic, Devonson4, Dioskorides, DivermanAU, Dj manton, Dodshe, Doesper, Double sharp, Drcola2, Electricmaster, Erik9, Flyer22 Reborn, Fotaun, Galoubet, General Epitaph, Good Olfactory, HJKeats, Headbomb, Hmrox, Ibro567, Imanderzel, JFG, Jack Gaines, Jameboy, Jamesx12345, Joefromrandb, JohnArmagh, JorisvS, Joseph Solis in Australia, Josve05a, Jrauser, KGyST, Keithpickering, LarplePelican, Leuko, Lgcharlot, Little green rosetta, Logan, Lusanaherandraton, Marc Kupper, Materialscientist, Mhym, Michael Hardy, Midnightdreary, Minimac's Clone, Modest Genius, MrBill3, Myasuda, NellieBly, Nergaal, Niceguyedc, Nimesh Mistry, NokiReblikaiTheVirus, NuclearWarfare, Oshwah, Pausch, Philroc, Plastikspork, Pleiotrop3, Pleonic, Quebec99, RJHall, RandomCritic, Remember, Rich Farmbrough, Richard Arthur Norton (1958-), Rjwilmsi, Roentgenium111, Ronstew, Ryoung122, SchfiftyThree, SchuminWeb, Seaphoto, Serendipodous, Shsilver, Simon Villeneuve, Skomorokh, Solomonfromfinland, Tamfang, Terry0051, Tom.Reding, Turgan, Twinsday, Viciouspiggy, VoABot II, Vsmith, Yamaguchi先生, Z0, 157 anonymous edits 29

Extraterrestrial diamonds *Source:* https://en.wikipedia.org/w/index.php?oldid=855935666 *License:* Creative Commons Attribution-Share Alike 3.0 *Contributors:* Alex Shih, Brandmeister, ClueBot NG, Cwmhiraeth, Dimadick, Georgelazenby, Grand-Duc, Headbomb, Huntster, J 1982, Kintetsubuffalo, Oshwah, Plantsurfer, Pomax, RockMagnetist (DCO visiting scholar), The Rambling Man, Tom.Reding, Yoninah, 13 anonymous edits 43

Kuiper belt *Source:* https://en.wikipedia.org/w/index.php?oldid=854370669 *License:* Creative Commons Attribution-Share Alike 3.0 *Contributors:* 2WR1, A2soup, ATX-NL, Affirst, Agmartin, Al Legorhythm, Al1549, Aliciamarie9313, Arjayay, Askari Mark, AwesomeAH18, BatteryIncluded, Bear-rings, Beland, Bender235, Bobdog54, ClueBot NG, Continentaleurope, Cygalder, Dan100, Dark Flow, Daviddaniel37, Dawnseeker2000, Dcirovic, DeadKing99, Deeday-UK, Doanri, DocWatson42, Double sharp, DrKay, EdChem, El C, Eric Kvaalen, Exoplanetaryscience, Favonian, Finnusertop, Fishyishere, Gadget850, Gah4, Gap9551, Gary D Robson, Gatemansgc, General Ization, Gob Lofa, Gts-tg, HamilcarBarca1, Hanif Al Husaini, Hazlogen, Hddty., Headbomb, Hibernian, Huntster, IronGargoyle, J 1982, JJMC89, JakeR, Jarble, Jehochman, Jiten D, Jmencisom, Jogar2, John, John "Hannibal" Smith, Jonesey95, JorisvS, Keith-264, Kheider, Kleinhern, Kleuske, Konlon15, KoshVorlon, Krj373, KylieTastic, LL221W, Laoris, Leschnei, Llywrch, Lugsciath, Master of Time, Materialscientist, Mild Bill Hiccup, MisfitToys, Modest Genius, Mrjulesd, Mzanon, Nardog, Natg 19, NewEnglandYankee, Nichodon, Oklo99, Omnipaedista, PoqVaUSA, ProloyAdhikary, ProprioMe OW, Quondum, RP88, Racerx11, Renerpho, Rfassbind, Ricardo A. Olea, RickinBaltimore, RioVerde, Rjwilmsi, Roentgenium111, Rowan Forest, Schaffman, SchutteGod, Serendipodous, Serols, Shellwood, Sir Cumference, Sleigh, Slightsmile, SpaceDude777, Stanning, Steinbach, Tarl N., Tbayboy, Tetra quark, Tom.Reding, Totallylegituser123, Trappist the monk, TwoTwoHello, Urhixidur, Wbm1058, Wieralee, Wiki Guy123098, Wikid77, Wikishovel, WilyD, WolfmanSF, ,62↑ 111 anonymous edits 49

Neptune trojan *Source:* https://en.wikipedia.org/w/index.php?oldid=856592312 *License:* Creative Commons Attribution-Share Alike 3.0 *Contributors:* 44Dume, Agmartin, Aldebaran66, Bender235, Blaxthos, Bryan Derksen, ChongDae, Danim, Double sharp, Dracontes, Dwarfplanets.org.uk, Eurocommuter, Exoplanetaryscience, Glagoli, Goustien, Headbomb, Hecseur, J 1982, Jan.Kamenicek, JojoTNO, JorisvS, Kheider, Kogge, Kuralyov, Kwamikagami, Ling.Nut, Mejor Los Indios, Mianreg, Nergaal, Njardarlogar, Ohms law, PraiseJezeus, RJHall, RSStockdale, Renerpho, Reyk, Rfassbind, Richardson mcphillips, Rjwilmsi, Rothorpe, Ruslik0, Sardanaphalus, Sedna17, SheriffIsInTown, Shipbegan, Simplexity22, Solonom, Starcluster, Starmaker it, Swordslayer18, TRBlom, Tamfang, The Singing Badger, The Tom, ThreeBlindMice, Tom.Reding, Tycho Magnetic Anomaly-1, Urhixidur, WolfmanSF, 18 anonymous edits 76

Formation and evolution of the Solar System *Source:* https://en.wikipedia.org/w/index.php?oldid=856485525 *License:* Creative Commons Attribution-Share Alike 3.0 *Contributors:* Agmartin, Anrnusna, Antoni Barau, Arbnos, Ashill, Astronaut Willis, Audin, Aveh8, BatteryIncluded, Bear-rings, Bender235, Benfxmth, Besselfunctions, BirdValiant, Blackninga3216, BlakePlays, Bomb319, CES1596, CLCStudent, CanadianLinuxUser, ChamithN, Charles Edwin Shipp, ClueBot NG, CyanoTex, Cyrus noto3at bulaga, DVdm, DatGuy, David.moreno72, Davros69999, Dawnseeker2000, Dbachmann, Dcirovic, Delphan Gruss EVI, DemocraticLuntz, Dodshe, Dolkungbighead, Download, DrKay, DrStrauss, Drbogdan, Dukebaker, Earthandmoon, Ehrenkater, El C, Eric Kvaalen, Escape Orbit, Excirial, FT2, Fama Clamosa, Fanta206, FlightTime, Friginator, Frinthruit, Gakrivas, Gap9551, GeneralizationsAreBad, GiantSnowman, Gilliam, Gilo1969, Hameltion, Hdjensofjfnen, Headbomb, Hillbillyholiday, Houjou, I dream of horses, Intel4004, Ira Leviton, Iridescent, IronGargoyle, J 1982, JFG, JaconaFrere, Jarble, Jcpag2012, Jim1138, Jmencisom, Joe Joe S.34567, John, JohnGolt, Jon Kolbert, JorisvS, Josve05a, Jwratner1, Kbh3rd, Kianu123456789123456789, Kostas20142, Kozmosis, L1A1 FAL, Lawrence5641, Lchappell, Lugia2453, MarkSewath, Materialscientist, Mchcopl, Messier8, MusikAnimal, NSH002, Newone, Niceguyedc, Niseto, Orenburg1, Oshwah, Oxcross, Pfalzner, PlanetUser, Quenhitran, RA0808, RP88, Reatlas, Red Planet X (Hercolubus), Reddi, Rjwilmsi, Rmosler2100, Rod57, Rothorpe, Ruslik0, Seaphoto, Serendipodous, Serols, Shellwood, Sir Cumference, Sithlord Timmy, Slightsmile, Space Infinite, Steven Corder, Tetra quark, Tfr000, The Herald, ToBeFree, Tom.Reding, Treisijs, Urhixidur, V620 Cephei, Vgy7ujm, Vsmith, WikiProffy2122, Wtmitchell, Xinxinliu, ZaperaWiki44, Zezen, 174 anonymous edits 83

Nice model *Source:* https://en.wikipedia.org/w/index.php?oldid=850138859 *License:* Creative Commons Attribution-Share Alike 3.0 *Contributors:* 216Kleopatra, A876, Agmartin, Ahpook, AshLin, Ashill, Assscroft, Belmanzoo, Bubba73, Charoniana, Citation bot 1, ClueBot NG, CuriousMind01, DePiep, Deflective, Długosz, Fences and windows, Fotaun, Ground Zero, Gulumeemee, Headbomb, Henrykus, I dream of horses, IJBall, IVAN3MAN, Igodard, Iridia, John, John of Reading, JorisvS, Kheider, Koavf, Kwamikagami, Loupeter, Messier8, MichaelSchoenitzer, Mike s, Mikek999, Mishac, Mortense, Nealmcb, Nick Moyes, R. S. Shaw, Reatlas, Rfassbind, Rjwilmsi, Roentgenium111, Rothorpe, Rursus, Rwflammang, Shuffdog, Silvrstridr, Smec, The Tom, Tom.Reding, Tony Mach, Urhixidur, Whoop whoop pull up, WolfmanSF, 23 anonymous edits 107

Moons of Neptune *Source:* https://en.wikipedia.org/w/index.php?oldid=855393522 *License:* Creative Commons Attribution-Share Alike 3.0 *Contributors:* 84user, AG Caesar, Anna Frodesiak, Anti-Quasar, Archon 2488, BilCat, Bubba73, Canterbury Tail, Casliber, ClueBot NG, Cnwilliams, Colonies Chris, DARTH SIDIOUS 2, Damonskye, Diannaa, Donner60, Double sharp, Doug Weller, Drhex, Emijrp, Exoplanetaryscience, Firsfron, Gadget850, Gap9551, Glacialfox, GoingBatty, Grafen, Graphium, Gsbs58, Headbomb, Hurricanefan24, J. 'mach' wust, Jarodalien, JavautilRandom, John of Reading, JorisvS, Kheider, Klilidiplomus, Kogge, Kwamikagami, Labtek00, Lanthanum-138, Lasunncty, Leaky caldron, Lk07162, MadGuy7023, Maile66, Marek69, Maxlaxp, Metiscus, Microchip08, MrBlueSky, MusikAnimal, Nergaal, Ninney, P. S. F. Freitas, Patteroast, PohranicniStraze, PureRED, Qzd, RandomCritic, Reatlas, Rich Farmbrough, Rjwilmsi, Roentgenium111, Rothorpe, Ruslik0, Scwlong, Seaphoto, Serendipodous, Siddiqsazzad001, Skywatcher68, Slightsmile, SoylentPurple, Stephenb, StewartIM, Tetra quark, The Rambling Man, Thincat, Tom.Reding, Tsinfandel, Tycho Magnetic Anomaly-1, Ucucha, Wasill37, Whoop whoop pull up, WolfmanSF, ZYjacklin, Zocke1r, 116 anonymous edits 117

Rings of Neptune *Source:* https://en.wikipedia.org/w/index.php?oldid=856480335 *License:* Creative Commons Attribution-Share Alike 3.0 *Contributors:* 84user, Alexikoua, Ariana-hime, Arsia Mons, Art LaPella, Asav, BOTarate, Begoon, Bender235, BilCat, BlueMoonlet, CapnZapp, Chmee2, Citation bot 1, ClueBot NG, ColonelKurtz2001, CompuHacker, Connormah, D.H, Danaman5, Dave Andrew, DavidLeighEllis, Dcirovic, Discospinster, Double sharp, DrKay, Earthandmoon, Elberth 00001939, Eragon845, Esrever, Eusebeus, Exoplanetaryscience, Fama Clamosa, Geeteshgadkari, General Epitaph, Gilliam, GrahamHardy, Headbomb, Hellbus, Howcheng, Icairns, J 1982, J.delanoy, JHunterJ, Ja 62, Joefromrandb, Jonathunder, Kaldari, Katydidit, Kogge, Leaky caldron, Lugia2453, MONGO, MarchOrDie, Materialscientist, Mattisse, Melonkelon, Mianreg, Michael Devore, Mild Bill Hiccup, Mitrius, Modest Genius, Moez, Nascar1996, Nihiltres, Nopira, PackMecEng, Pbouchet, Peter Horn, Peter.C, Philip Trueman, Piledhigheranddeeper, Plastikspork, Polarscribe, Protargol, Quackjack29, RJHall, Randy Kryn, Reatlas, RedSirus, Renaissancee, Reywas92, Rfassbind, Rich Farmbrough, Rjwilmsi, RobertG, Robinh, Rorschach, Rothorpe, Ruslik0, SandyGeorgia, Seaphoto, Serendipodous, Simoncursitor, Spitfire, Steel1943, Stephen B Streater, Surajt88, Tbhotch, Thatguyflint, Tom.Reding, Tommy2010, Trackteur, Trappist the monk, Treisijs, Tycho Magnetic Anomaly-1, WereSpielChequers, Whoop whoop pull up, Wikid77, Wikipelli, WolfmanSF, Xession, 97 anonymous edits 131

Exploration of Neptune *Source:* https://en.wikipedia.org/w/index.php?oldid=854829866 *License:* Creative Commons Attribution-Share Alike 3.0 *Contributors:* 72, Altzinn, Andy120290, Arjayay, BatteryIncluded, Boaex, BranStark, ChiZeroOne, ClueBot NG, Cmglee, DN-boards1, DPBT1, DangerousJXD, Dfoofnik, Double sharp, Dr.Toonhattan, EdChem, Ehrenkater, Eyreland, FoCuSandLeArN, Gilliam, Goudzovski, Harlock81, Headbomb, Hillbillyholiday, I dream of horses, IanOsgood, JFG, Jonatan Swift, JorisvS, Krapenhoeffer, LarplePelican, Me, Myself, and I are Here, Mild Bill Hiccup, Myasuda, Noob person, Northgrove, OlEnglish, Perfection, Randy Kryn, Rcsprinter123, RhinoMind, RobDe68, Ruslik0, Sam Sailor, Shrewpelt, Tamfang, Teamcoltra, Tom.Reding, Tomruen, Triton66, Truthanado, 28 anonymous edits 143

Image Sources, Licenses and Contributors

The sources listed for each image provide more detailed licensing information including the copyright status, the copyright owner, and the license conditions.

Image *Source:* https://en.wikipedia.org/w/index.php?title=File:Padlock-silver.svg *Contributors:* AzaToth, BotMultichill, BotMultichillT, Gurch, Jarekt, Kallerna, Multichill, Perhelion, Rd232, Riana, Sarang, Siebrand, Steinsplitter, 4 anonymous edits 1
Image *Source:* https://en.wikipedia.org/w/index.php?title=File:Neptune_Full.jpg *License:* Public Domain *Contributors:* Alloo02, ArionEstar, Arjuno3, Armbrust, Brateevsky, Bulat, Clarice Reis, Coriolis ende, Danim, Fabio Bettani, Huntster, Ischa1, J 1982, Jarekt, Jcpag2012, Josve05a, Körnerbrötchen, Li-sung, Moheen Reeyad, Nagualdesign, Newone, O'Dea, Orange-kun, PlanetUser, Quenhitran, Ruslik0, Str4nd, TheFreeWorld, Z7504, باسل, 3 anonymous edits 1
Image *Source:* https://en.wikipedia.org/w/index.php?title=File:Loudspeaker.svg *License:* Public Domain *Contributors:* User:Dbenbenn, User:Optimager, User:Tsca, User:Dbenbenn, User:Optimager, User:Tsca, User:Dbenbenn, User:Optimager, User:Tsca 1
Figure 1 *Source:* https://en.wikipedia.org/w/index.php?title=File:Galileo.arp.300pix.jpg *License:* Public Domain *Contributors:* ABF, Alefisico, Allforrous, Alno, BotMultichill, BotMultichillT, Bukk, Daniele Pugliesi, David J Wilson, Deadstar, Dirk Hünniger, G.dallorto, Gary King, Herbythyme, Huntster, Kam Solusar, Liberal Freemason, Michael Bednarek, Phrood∼commonswiki, Pérez∼commonswiki, Quadell, Ragesoss, Schaengel89∼commonswiki, Semnoz, Shakko, Silverije, Trijnstel, Túrelio, Un1c0s bot∼commonswiki, Yonatanh, Ángel el Astrónomo, 28 anonymous edits 5
Figure 2 *Source:* https://en.wikipedia.org/w/index.php?title=File:Urbain_Le_Verrier.jpg *License:* Public Domain *Contributors:* User Magnus Manske on en.wikipedia 6
Figure 3 *Source:* https://en.wikipedia.org/w/index.php?title=File:Neptune,_Earth_size_comparison_2.jpg *License:* Public Domain *Contributors:* User:Jcpag2012 8
Figure 4 *Source:* https://en.wikipedia.org/w/index.php?title=File:Neptune_diagram.svg *License:* Creative Commons Attribution-Sharealike 3.0 *Contributors:* NASA; Pbroks13 (redraw) 9
Figure 5 *Source:* https://en.wikipedia.org/w/index.php?title=File:Neptune-Methane.jpg *License:* Public domain *Contributors:* Hubble Space Telescope 10
Figure 6 *Source:* https://en.wikipedia.org/w/index.php?title=File:Neptune's_Dynamic_Environment.webm *License:* Public Domain *Contributors:* 1Veertje, Montvale, Reguyla, Ruslik0 11
Figure 7 *Source:* https://en.wikipedia.org/w/index.php?title=File:Neptune_clouds.jpg *License:* Public Domain *Contributors:* NASA / Jet Propulsion Lab 12
Figure 8 *Source:* https://en.wikipedia.org/w/index.php?title=File:Neptune_storms.jpg *License:* Public Domain *Contributors:* NASA/Voyager 2 Team 14
Figure 9 *Source:* https://en.wikipedia.org/w/index.php?title=File:Neptune's_Great_Dark_Spot.jpg *License:* Public Domain *Contributors:* NASA / Jet Propulsion Lab 15
Figure 10 *Source:* https://en.wikipedia.org/w/index.php?title=File:Neptune's_shrinking_vortex.jpg *Contributors:* Huntster, Jmencisom 16
Figure 11 *Source:* https://en.wikipedia.org/w/index.php?title=File:Different_Faces_Neptune.jpg *Contributors:* - 17
Figure 12 *Source:* https://en.wikipedia.org/w/index.php?title=File:Neptune_Orbit.gif *License:* Creative Commons Attribution-Sharealike 3.0 *Contributors:* User:Lookang 18
Figure 13 *Source:* https://en.wikipedia.org/w/index.php?title=File:TheKuiperBelt_classes-en.svg *License:* Creative Commons Attribution-Sharealike 2.5 *Contributors:* Lilyu and Eurocommuter 19
Figure 14 *Source:* https://en.wikipedia.org/w/index.php?title=File:Lhborbits.png *License:* Creative Commons Attribution-Sharealike 3.0,2.5,2.0,1.0 *Contributors:* en:User:AstroMark 20
Figure 15 *Source:* https://en.wikipedia.org/w/index.php?title=File:Neptune-visible.jpg *License:* Public Domain *Contributors:* Jcpag2012, Jianhui67, Phrood∼commonswiki, Ruslik0, 6 anonymous edits 21
Figure 16 *Source:* https://en.wikipedia.org/w/index.php?title=File:Proteus_(Voyager_2).jpg *License:* Public Domain *Contributors:* Voyager 2, NASA 22
Figure 17 *Source:* https://en.wikipedia.org/w/index.php?title=File:Neptunerings.jpg *License:* Public Domain *Contributors:* Cocu, Jameslwoodward, Magog the Ogre, OgreBot 2, PlanetUser, Stefan2, Torsch 23
Figure 18 *Source:* https://en.wikipedia.org/w/index.php?title=File:Neptune_from_the_VLT_with_MUSE_GALACSI_Narrow_Field_Mode_adaptive_optics.jpg *Contributors:* Huntster, Sergkarman 24
Figure 19 *Source:* https://en.wikipedia.org/w/index.php?title=File:Triton_moon_mosaic_Voyager_2_(large).jpg *License:* Public Domain *Contributors:* NASA / Jet Propulsion Lab / U.S. Geological Survey 25
Image *Source:* https://en.wikipedia.org/w/index.php?title=File:Cscr-featured.svg *License:* GNU Lesser General Public License *Contributors:* Anomie 27
Figure 20 *Source:* https://en.wikipedia.org/w/index.php?title=File:Sternwarte_Berlin_Schinkel.jpg *License:* Public Domain *Contributors:* Dr.michi, Infrogmation, Lotse, Pilettes, SSagman 30
Figure 21 *Source:* https://en.wikipedia.org/w/index.php?title=File:Gravitational_perturbation.svg *License:* Creative Commons Attribution-ShareAlike 3.0 Unported *Contributors:* RJHall 32
Figure 22 *Source:* https://en.wikipedia.org/w/index.php?title=File:John_Couch_Adams.jpg *License:* Public Domain *Contributors:* GDK, Ra'ike, Wutsje 33
Figure 23 *Source:* https://en.wikipedia.org/w/index.php?title=File:Urbain_Le_Verrier.jpg *License:* Public Domain *Contributors:* User Magnus Manske on en.wikipedia 34
Figure 24 *Source:* https://en.wikipedia.org/w/index.php?title=File:Johann-Gottfried-Galle.jpg *License:* Public Domain *Contributors:* (unknown) 35
Image *Source:* https://en.wikipedia.org/w/index.php?title=File:Neptune_discovery.png *License:* GNU Free Documentation License *Contributors:* Bryan Derksen, Keithpickering, Sfan00 IMG 36
Figure 25 *Source:* https://en.wikipedia.org/w/index.php?title=File:Neptune_Full.jpg *License:* Public Domain *Contributors:* Alloo02, ArionEstar, Arjuno3, Armbrust, Brateevsky, Bulat, Clarice Reis, Coriolis ende, Danim, Fabio Bettani, Huntster, Ischa1, J 1982, Jarekt, Jcpag2012, Josve05a, Körnerbrötchen, Li-sung, Moheen Reeyad, Nagualdesign, Newone, O'Dea, Orange-kun, PlanetUser, Quenhitran, Ruslik0, Str4nd, TheFreeWorld, Z7504, باسل, 3 anonymous edits 38
Figure 26 *Source:* https://en.wikipedia.org/w/index.php?title=File:SpaceNanoDiamonds.png *License:* Public Domain *Contributors:* RockMagnetist (DCO visiting scholar) 44
Figure 27 *Source:* https://en.wikipedia.org/w/index.php?title=File:Uranus_(Edited).png *License:* Public Domain *Contributors:* Jcpag2012, PlanetUser 46
Figure 28 *Source:* https://en.wikipedia.org/w/index.php?title=File:Moissanite-USGS-20-1002a.jpg *License:* Public Domain *Contributors:* Andrew Silver 47
Figure 29 *Source:* https://en.wikipedia.org/w/index.php?title=File:Kuiper_belt_plot_objects_of_outer_solar_system.png *License:* Creative Commons Attribution-Sharealike 3.0 *Contributors:* Jeanjung212, JorisvS, Mackensen, NURUL1234, OgreBot 2, Rfassbind, Sprachpflegger 50
Figure 30 *Source:* https://en.wikipedia.org/w/index.php?title=File:Gerard_Kuiper_1964b.jpg *Contributors:* Gelderen, Hugo van / Anefo 52
Figure 31 *Source:* https://en.wikipedia.org/w/index.php?title=File:Maunatele.jpg *License:* Public Domain *Contributors:* NASA 54
Figure 32 *Source:* https://en.wikipedia.org/w/index.php?title=File:Dust_Models_Paint_Alien's_View_of_Solar_System.ogv *License:* Public Domain *Contributors:* NASA 55
Figure 33 *Source:* https://en.wikipedia.org/w/index.php?title=File:KBOs_and_resonances.png *Contributors:* User:Renerpho 57
Figure 34 *Source:* https://en.wikipedia.org/w/index.php?title=File:TheKuiperBelt_classes-en.svg *License:* Creative Commons Attribution-Sharealike 2.5 *Contributors:* Lilyu and Eurocommuter 57
Figure 35 *Source:* https://en.wikipedia.org/w/index.php?title=File:Hot_and_cold_KBO.svg *Contributors:* Doanri 59
Figure 36 *Source:* https://en.wikipedia.org/w/index.php?title=File:Lhborbits.png *License:* Creative Commons Attribution-Sharealike 3.0,2.5,2.0,1.0 *Contributors:* en:User:AstroMark 60
Figure 37 *Source:* https://en.wikipedia.org/w/index.php?title=File:2003_UB313_near-infrared_spectrum.png *License:* GNU Free Documentation License *Contributors:* 1989, GifTagger, Jeanjung212, Steinsplitter, User156473 62
Figure 38 *Source:* https://en.wikipedia.org/w/index.php?title=File:Artist's_impression_of_exiled_asteroid_2004_EW95.jpg *Contributors:* Jmencisom 63

Figure 39 *Source:* https://en.wikipedia.org/w/index.php?title=File:TheKuiperBelt_PowerLaw2.svg *License:* Creative Commons Attribution-ShareAlike 3.0 Unported *Contributors:* User:Eurocommuter 65
Figure 40 *Source:* https://en.wikipedia.org/w/index.php?title=File:TheKuiperBelt_Projections_100AU_Classical_SDO.svg *License:* Creative Commons Attribution-ShareAlike 3.0 Unported *Contributors:* User:Eurocommuter∼commonswiki 67
Figure 41 *Source:* https://en.wikipedia.org/w/index.php?title=File:Triton_moon_mosaic_Voyager_2_(large).jpg *License:* Public Domain *Contributors:* NASA / Jet Propulsion Lab / U.S. Geological Survey 68
Figure 42 *Source:* https://en.wikipedia.org/w/index.php?title=File:EightTNOs.png *License:* Creative Commons Attribution-ShareAlike 3.0 Unported *Contributors:* Lexicon 69
Figure 43 *Source:* https://en.wikipedia.org/w/index.php?title=File:14-281-KuiperBeltObject-ArtistsConcept-20141015.jpg *License:* Public Domain *Contributors:* Drbogdan 71
Figure 44 *Source:* https://en.wikipedia.org/w/index.php?title=File:KBO_2014_MU69_HST.jpg *License:* Public domain *Contributors:* Bodhisattwa, Huntster, LL221W, PhilipTerryGraham, Rfassbind, Ruslik0, Wieralee, Ziembo 72
Figure 45 *Source:* https://en.wikipedia.org/w/index.php?title=File:2014_MU69_orbit.jpg *License:* Public Domain *Contributors:* DmitTrix, Gazebo, Huntster, LL221W, PhilipTerryGraham, Torsch, Wieralee, Yann 72
Figure 46 *Source:* https://en.wikipedia.org/w/index.php?title=File:Kuiper_belt_remote.jpg *License:* Public domain *Contributors:* User:Audriusa 74
Image *Source:* https://en.wikipedia.org/w/index.php?title=File:Commons-logo.svg *License:* logo *Contributors:* Anomie, Callanecc, CambridgeBayWeather, Jo-Jo Eumerus, RHaworth 75
Figure 47 *Source:* https://en.wikipedia.org/w/index.php?title=File:NTrojans_Plutinos_55AU.svg *License:* Creative Commons Attribution-Sharealike 3.0 *Contributors:* Eurocommuter 77
Figure 48 *Source:* https://en.wikipedia.org/w/index.php?title=File:Neptunian_Trojans.gif *Contributors:* frankuitaalst from the Gravity Simulator message board. 78
Image *Source:* https://en.wikipedia.org/w/index.php?title=File:Lock-green.svg *License:* Creative Commons Zero *Contributors:* User:Trappist the monk 81
Figure 49 *Source:* https://en.wikipedia.org/w/index.php?title=File:Protoplanetary-disk.jpg *License:* Public Domain *Contributors:* NASA 84
Figure 50 *Source:* https://en.wikipedia.org/w/index.php?title=File:Pierre-Simon_Laplace.jpg *License:* Public Domain *Contributors:* Ashill, Ecummenic, Elcobbola, Gene.arboit, Jimmy44, Laura1822, Leyo, Mattes, NicoScribe, Olivier, 霧木諒二 , 2 anonymous edits 85
Figure 51 *Source:* https://en.wikipedia.org/w/index.php?title=File:M42proplyds.jpg *License:* Public domain *Contributors:* C.R. O'Dell/Rice University; NASA 86
Figure 52 *Source:* https://en.wikipedia.org/w/index.php?title=File:Solarnebula.jpg *License:* Public Domain *Contributors:* William K. Hartmann, Planetary Science Institute, Tucson, Arizona UNIQ-ref-0-0235c40ef49a14df-QINU 88
Figure 53 *Source:* https://en.wikipedia.org/w/index.php?title=File:Artist's_concept_of_collision_at_HD_172555.jpg *License:* Public Domain *Contributors:* NASA/JPL-Caltech 90
Figure 54 *Source:* https://en.wikipedia.org/w/index.php?title=File:Lhborbits.png *License:* Creative Commons Attribution-Sharealike 3.0,2.5,2.0,1.0 *Contributors:* en:User:AstroMark 92
Image *Source:* https://en.wikipedia.org/w/index.php?title=File:Interactive_icon.svg *Contributors:* User:Evolution and evolvability 95
Figure 55 *Source:* https://en.wikipedia.org/w/index.php?title=File:Barringer_Crater_aerial_photo_by_USGS.jpg *License:* Public Domain *Contributors:* USGS/D. Roddy 95
Figure 56 *Source:* https://en.wikipedia.org/w/index.php?title=File:Voyager_2_Neptune_and_Triton.jpg *License:* Public Domain *Contributors:* NASA / Jet Propulsion Lab 99
Figure 57 *Source:* https://en.wikipedia.org/w/index.php?title=File:Sun_red_giant.svg *License:* GNU Free Documentation License *Contributors:* Oona Räisänen (User:Mysid), User:Mrsanitazier. 100
Figure 58 *Source:* https://en.wikipedia.org/w/index.php?title=File:M57_The_Ring_Nebula.JPG *License:* Public domain *Contributors:* The Hubble Heritage Team (AURA/STScI/NASA) 102
Figure 59 *Source:* https://en.wikipedia.org/w/index.php?title=File:Milky_Way_Spiral_Arm.svg *License:* GNU Free Documentation License *Contributors:* User:Surachit 103
Image *Source:* https://en.wikipedia.org/w/index.php?title=File:Solar_Life_Cycle.svg *License:* Public Domain *Contributors:* Oliverbeatson 105
Image *Source:* https://en.wikipedia.org/w/index.php?title=File:Timeline_icon.svg *License:* Creative Commons Attribution-ShareAlike 3.0 Unported *Contributors:* User:Dschwen 105
Figure 60 *Source:* https://en.wikipedia.org/w/index.php?title=File:Lhborbits.png *License:* Creative Commons Attribution-Sharealike 3.0,2.5,2.0,1.0 *Contributors:* en:User:AstroMark 108
Figure 61 *Source:* https://en.wikipedia.org/w/index.php?title=File:Tsiganis2005-1.svg *Contributors:* Tsiganis et al. 110
Figure 62 *Source:* https://en.wikipedia.org/w/index.php?title=File:Voyager_2_Neptune_and_Triton.jpg *License:* Public Domain *Contributors:* NASA / Jet Propulsion Lab 118
Figure 63 *Source:* https://en.wikipedia.org/w/index.php?title=File:Tritonian_sky.jpg *License:* Public domain *Contributors:* ComputerHotline, Huntster, Liftarn, OgreBot 2, Ruslik0, Themightyquill, Tomruen, XBrain130, 1 anonymous edits 119
Figure 64 *Source:* https://en.wikipedia.org/w/index.php?title=File:Outer_planet_moons.png *Contributors:* User:StewartIM 119
Figure 65 *Source:* https://en.wikipedia.org/w/index.php?title=File:Proteus_3D_model.ogv *License:* Creative Commons Attribution-Sharealike 3.0 *Contributors:* Ruslik0 121
Figure 66 *Source:* https://en.wikipedia.org/w/index.php?title=File:Neptune's_Dynamic_Environment.webm *License:* Public Domain *Contributors:* 1Veertje, Montvale, Reguyla, Ruslik0 121
Figure 67 *Source:* https://en.wikipedia.org/w/index.php?title=File:TheIrregulars_NEPTUNE.svg *License:* Creative Commons Attribution-ShareAlike 3.0 Unported *Contributors:* User:Eurocommuter 123
Figure 68 *Source:* https://en.wikipedia.org/w/index.php?title=File:Triton_orbit_&_Neptune.png *License:* Public Domain *Contributors:* User: ZYjacklin 124
Figure 69 *Source:* https://en.wikipedia.org/w/index.php?title=File:Masa_de_triton.svg *License:* Creative Commons Zero *Contributors:* AG Caesar, Capmo, Sarang 125
Image *Source:* https://en.wikipedia.org/w/index.php?title=File:Naiad_Voyager.png *License:* Public Domain *Contributors:* NASA 127
Image *Source:* https://en.wikipedia.org/w/index.php?title=File:Despina.jpg *License:* Public Domain *Contributors:* NASA 127
Image *Source:* https://en.wikipedia.org/w/index.php?title=File:Galatea_moon.jpg *License:* Public Domain *Contributors:* Voyager 2 127
Image *Source:* https://en.wikipedia.org/w/index.php?title=File:Larissa_1.jpg *License:* Public Domain *Contributors:* NASA/JPL/Space Science Institute 127
Image *Source:* https://en.wikipedia.org/w/index.php?title=File:S-2004_N1_Hubble_montage.jpg *License:* Public domain *Contributors:* Alan, Hibernian, Huntster, Jacopo Werther, Jcpag2012, JuTa, Jérémy-Günther-Heinz Jähnick, Lasunncty, SenseiAC, WolfmanSF 127
Image *Source:* https://en.wikipedia.org/w/index.php?title=File:Nereid-Voyager2.jpg *License:* Public Domain *Contributors:* Argo Navis, Augiasstallputzer∼commonswiki, ComputerHotline, Filemon, Leyo, Li-sung, PlanetUser, Ruslik0, Yarl 128
Image *Source:* https://en.wikipedia.org/w/index.php?title=File:Halimede.jpg *License:* Public Domain *Contributors:* NASA 128
Image *Source:* https://en.wikipedia.org/w/index.php?title=File:Psamathe_arrow.png *License:* Public Domain *Contributors:* Exoplanetaryscience 128
Figure 70 *Source:* https://en.wikipedia.org/w/index.php?title=File:Neptunian_rings_scheme_2.svg *License:* Creative Commons Attribution-Sharealike 3.0 *Contributors:* User:Fama Clamosa 132
Figure 71 *Source:* https://en.wikipedia.org/w/index.php?title=File:PIA02202_Neptune's_full_rings.jpg *License:* Public Domain *Contributors:* NASA/Jet Propulsion Lab 133
Figure 72 *Source:* https://en.wikipedia.org/w/index.php?title=File:PIA02224_Neptune's_rings.jpg *License:* Public Domain *Contributors:* NASA/Jet Propulsion Lab 134
Figure 73 *Source:* https://en.wikipedia.org/w/index.php?title=File:Neptune_ring_arcs.jpg *License:* Public Domain *Contributors:* NASA/Voyager 2 Team 136
Figure 74 *Source:* https://en.wikipedia.org/w/index.php?title=File:Neptune.jpg *License:* Public Domain *Contributors:* NASA/JPL 144
Figure 75 *Source:* https://en.wikipedia.org/w/index.php?title=File:Triton_moon_mosaic_Voyager_2_(large).jpg *License:* Public Domain *Contributors:* NASA / Jet Propulsion Lab / U.S. Geological Survey 144
Figure 76 *Source:* https://en.wikipedia.org/w/index.php?title=File:Voyager.jpg *License:* Public Domain *Contributors:* NASA 145
Figure 77 *Source:* https://en.wikipedia.org/w/index.php?title=File:Proteus_(Voyager_2).jpg *License:* Public Domain *Contributors:* Voyager 2, NASA 146

License

Creative Commons Attribution-Share Alike 3.0
//creativecommons.org/licenses/by-sa/3.0/

Index

(120216) 2004 EW95, 63
(15760) 1992 QB1, 56
(181708) 1993 FW, 54
(19308) 1996 TO66, 64
(309239) 2007 RW10, 19, 81
(316179) 2010 EN65, 81
(385695) 2005 TO74, 80

Abiogenesis, 106
Absolute magnitude, 66, 80
Absorption line, 62
Abundance of the chemical elements, 89
Académie des sciences, 34
Academy of the Hebrew Language, 7
Acceleration, 2
Accretion (astrophysics), 20, 87
Acetylene, 12, 14, 64
Active galactic nucleus, 104
Adams ring, 120
Adaptive optics, 4, 24, 148
Adrastea (moon), 99
Alastair GW Cameron, 52
Albedo, 3, 14, 56, 63, 66, 134, 155
Albert Einstein, 85
Alexis Bouvard, 3, 5, 31
Almahata Sitta, 45
Almond, 145
Aluminium, 88
Amazons, 79
Ammonia, 3, 4, 9, 49, 63, 64
Ammonium hydrosulfide, 3
Ammonium sulfide, 11
Amorphous, 47
André Brahic, 131
Andromeda–Milky Way collision, 106
Andromeda Galaxy, 106
Angular diameter, 3, 24
Angular momentum, 84, 88, 98, 108
Anticyclone, 15, 145
Aphelion, 80
Apoapsis, 21
Apparent magnitude, 3, 23, 30, 67
Apparent retrograde motion, 4, 25
Applied Physics Laboratory, 73

Apsides, 156
Apsis, 54, 123
Arc (geometry), 120
Arcsecond, 24
Argo (spacecraft), 26, 148
Argument of periapsis, 2
Aristarchus of Samos, 84
Arizona, 95
Armin Otto Leuschner, 51
Arthur Stanley Eddington, 85
Artifact (observational), 132
ArXiv, 81
Asteroid, 23, 70, 83, 97, 120
Asteroid belt, 18, 49, 56, 91, 111, 153
Asteroid capture, 53
Astronomer, 51
Astronomer Royal, 5, 29, 33
Astronomical naming conventions, 80
Astronomical symbol, 3
Astronomical unit, 1, 3, 17, 49, 50, 80, 85, 87, 102, 119, 148
Astronomical units, 108
Astronomy, 83, 109
Astronomy Cast, 27, 75
Asymptotic giant branch, 101, 106
Athenaeum (British magazine), 33
Atmosphere, 10, 123, 143
Atmosphere of Earth, 44
Atmospheric absorption, 148
Atmospheric composition, 3
Atmospheric methane, 10
Atmospheric pressure, 16
Atom Probe Tomography (APT), 44
Auroras, 145
Axial tilt, 2, 14, 97

B2FH paper, 85
Backscatter, 138
Bar (unit), 10, 16, 123
Barycenter in astronomy, 17
Barycentric coordinates (astronomy), 17
Berlin, 35, 46
Berlin Observatory, 6, 29, 35
Bibcode, 39–41, 81, 107

Big Bang nucleosynthesis, 86
Big Crunch, 104
Big Rip, 104
Binary asteroid, 126
Black dwarf, 102
Blink comparator, 52, 53
Blue giant, 104
Bodes law, 32, 37
Boeing, 148
Bond albedo, 3, 123, 134, 138
Book of Psalms, 7
Bow shock, 13
BPM 37093, 48
Brady Haran, 27
Brian G. Marsden, 55
Bureau des Longitudes, 7
Burgess, 151
Bya, 83, 105, 106

Calcium–aluminium-rich inclusion, 86
Cambridge, 29
Cambridge Observatory, 5, 33
Camille Flammarion, 120
Canyon Diablo meteorite, 105
Carbon, 101, 102
Carbon-12, 44
Carbon-13, 44
Carbonado, 47
Carbon dioxide, 12, 123
Carbon monoxide, 12, 64, 69, 123
Carme group, 125
Carnegie Institution for Science, 48
Cassini–Huygens, 26, 99, 147
Celsius, 4
Centaur (minor planet), 49, 67, 76, 78, 79, 119
Centaur (planetoid), 52
Centaurus, 48
Ceres (dwarf planet), 23, 70, 111
Cerro Tololo Inter-American Observatory, 53
Chaos theory, 97
Charge-coupled device, 54
Charles Kowal, 52
Charon (moon), 50, 60, 97, 155
Chelyabinsk meteor, 95
Chile, 38
Circumstellar disc, 49
Circumstellar habitable zone, 100
Cis-Neptunian object, 49, 76
CITEREFDuncanLissauer1997, 154
CITEREFRandall2015, 153
CITEREFZeilikGregory1998, 154
Classical Kuiper belt, 71
Classical Kuiper belt object, 19, 49, 56, 76, 79
Clearing the neighborhood, 70
Clock drive, 39
Clovis culture, 45

Clyde Tombaugh, 51, 52, 55
Collisional family, 112
Columbia University Press, 26
Comet, 51, 79, 92, 93, 96
Comet Hale–Bopp, 53
Comet Shoemaker–Levy 9, 95
Comet Wild 2, 89
Committee on Small Body Nomenclature, 56
Commons:Category:Kuiper belt, 75
Commons:Category:Moons of Neptune, 129
Commons:Category:Nice Model, 116
Commons:Category:Rings of Neptune, 141
Compound (chemistry), 62
Concave mirrors, 122
Conjunction (astronomy), 30
Conjunction (astronomy and astrology), 4
Conservation of angular momentum, 87
Convection, 13, 45
Co-orbital configuration, 78
Copley medal, 29, 36
Coronograph, 74
Cosmic dust, 55, 131, 134
Cosmic microwave background, 133
Cosmic ray, 64
CRC Press, 26
Cross section (geometry), 155
Cryovolcano, 73, 123
C-type asteroid, 63
Cubewano, 57
Cubewanos, 56
Cumulative distribution function, 104
Cyan, 10

Daniel Jones (phonetician), 155
Dark Matter and the Dinosaurs, 74
David Jewitt, 53
David J. Tholen, 118
Day, 127
Debris disc, 74
Declination, 3
Definition of planet, 70
Degenerate matter, 102
Degree (angle), 4, 80, 127
Deimos (moon), 96
Dennis Rawlins, 6
Density, 2
Desdemona (moon), 99
Despina (moon), 10, 21, 119, 120, 127, 131, 134, 135, 146
Detached object, 49, 76, 87
Deutsches Museum, 39
Diameter, 80
Diamond, 43
Diamond anvil cell, 45
Diamondoid, 45
Dictionary of Scientific Biography, 39, 40

Differential rotation, 18
Diffuse reflection, 155
Digital object identifier, 40, 41, 75, 81, 107
Dirty snowball, 52
Discovery of Neptune, **29**, 118
Discovery of Pluto, 8
Distant minor planet, 49, 76
Drag (physics), 88
D-type asteroid, 111
Dwarf planet, 8, 21, 23, 49, 70, 76
Dynamical friction, 109
Dynamical simulation, 107
Dynamo, 13

Earth, 3, 43, 44, 84, 88, 98, 101, 105, 143
Earth mass, 65, 88, 89
Earths mantle, 44
Earths rotation, 39
Ecliptic, 2
Edward Guinan, 22
Electrical conductor, 13
Electromagnetic radiation, 155
Electromagnetic spectrum, 148
Electronic computers, 32
Element (chemistry), 62
Ellipse, 97
Ellipsoid, 146
Elliptical galaxy, 104
El:Ποσειδώνας (πλανήτης), 151
Emanuel Swedenborg, 83, 84
Empirical observation, 3
Encyclopædia Britannica, 39, 152
English, 1, 49, 107, 127, 128
Epoch (astronomy), 1, 50
Equator, 2, 8
Equatorial mount, 34, 39
Eris (dwarf planet), 50, 68, 70
Erosion, 105
Errors and residuals in statistics, 32
ESA, 143, 147
Escape velocity, 2
Escarpment, 123
Ethane, 3, 12, 64
Ethylene, 64
Europa Clipper, 147
Europa (moon), 96
European Southern Observatory, 24
Exosphere, 11
Expected value, 66
Exploration of Neptune, **143**
Extrasolar planet, 9
Extrasolar planets, 83
Extraterrestrial diamonds, **43**
Extraterrestrial skies, 119

Fahrenheit, 4

Final examination, 32
Five-planet Nice model, 61
Fixed star, 4, 30
Fixed stars, 5
Flagship Program, 26, 147
Flattening, 2
Formation and evolution of the Solar System, 43, 49, 65, **83**, 107
Formation of the Kuiper belt, 61
Forward scatter, 138
Fossil record, 103
François Arago, 7, 29, 37, 131, 135
Frederick C. Leonard, 51
Fred Hoyle, 85
Fred Whipple, 52, 55
French Republic, 134
French Revolution, 134
Friedrich Georg Wilhelm von Struve, 7
Fritz Haber Institute of the Max Planck Society, 46
Frost line (astrophysics), 88
Fullerene, 44

Galactic Center, 76, 77
Galactic Centre, 103
Galactic tide, 96, 103
Galactic year, 103
Galatea (moon), 10, 21, 23, 119, 120, 127, 131, 134, 136, 138, 146
Galilean moons, 23, 98
Galileo Galilei, 4, 5, 30
Galileo (spacecraft), 147
Ganymede (moon), 96
Gas giant, 8
Gauss (unit), 13
Geographical pole, 2
Geological consequences on Earth, 106
Geologic time scale, 95
Geology, 83
Geometric albedo, 3, 122, 123
George Airy, 29
George Biddell Airy, 5, 33, 39
George Phillips Bond, 155
Gerard Kuiper, 50–52, 118, 128
Geyser, 123
Geysers, 146
G-force, 2
Giant impact hypothesis, 83, 90
Giant-impact hypothesis, 97
Giant planet, 3, 8, 88, 107, 108, 122, 125
Giant star, 43
Gigametre, 123
Gold Medal of the Royal Astronomical Society, 37
Graben, 123
Grand tack hypothesis, 93

Graphical timeline of the universe, 105
Graphite, 43, 44
Gravitation, 3, 5
Gravitational collapse, 83, 126
Gravitational drag, 91
Gravity, 8, 104, 146
Gravity assist, 108, 148
Gravity wave, 12
Great Dark Spot, 1, 4, 14, 15, 145
Great Red Spot, 4, 15, 145
Greek language, 7
Greek mythology, 7, 79, 117, 120
Greenhouse gas, 100
G Ring, 137
Ground-based telescope, 4

Habitable zone, 106
Halimede (moon), 119, 122, 128
Halleys Comet, 53
Harold F. Levison, 108
Harold J. Reitsema, 132
Haumea, 49, 69, 70, 73
HD 139664, 74
HD 53143, 74
Heat, 45, 87
Hebrew language, 7
Heinrich Louis dArrest, 6, 29, 35
Heliocentric, 17
Heliocentrism, 84
Heliosphere, 73
Helium, 3, 4, 10, 86
Helium flash, 101
Help:IPA for English, 127
Herschel Space Observatory, 124
Hertzsprung–Russell diagram, 97, 106
Hilda asteroids, 111
Hills cloud, 51
Hill sphere, 123, 125
Hindu astrology, 7
Horizontal branch, 101, 106
Horseshoe orbit, 78
Hubble Space Telescope, 4, 15, 21, 24, 63, 74, 87, 133, 143
Hydrate, 64
Hydrocarbon, 4, 64
Hydrogen, 3, 4, 10, 86
Hydrogen cyanide, 12
Hydrogen deuteride, 3
Hydrogen ion, 10
Hydrogen sulfide, 11, 64
Hydrostatic equilibrium, 87, 117
Hyperion (moon), 124

IAU, 56
Ice giant, 4, 8, 61, 73, 92, 109, 112
Immanuel Kant, 83, 84

Impact event, 109
Inch, 39
Inclination, 80
Inclusion (mineral), 45
Infrared, 10, 25, 55, 137
Infrared excess, 74
Infrared Space Observatory, 45
Inner rings, 120
Inner Solar System, 87, 107
In situ, 122
Internal heating, 16
Internal structure, 9
International Astronomical Union, 8, 50, 70, 79, 80
International Journal of Astrobiology, 81
International Standard Book Number, 26, 39, 40, 74, 107, 155
Interstellar comet, 154
Interstellar medium, 43, 73, 86, 96, 105
Interstellar space, 53, 84
Invariable plane, 2, 151
Inverse problem, 32
Io (moon), 96
Iron, 88
Iron-60, 86
Irregular moon, 123
Irregular satellite, 112, 117
Isotopic signature, 43

J2000, 1, 151
James Challis, 5, 33
James Glaisher, 36
Jane Luu, 53
Jan Oort, 53
Janus (mythology), 7
Jérôme Lalande, 30, 31
Jet Propulsion Laboratory, 143
Jewish folklore, 7
Johann Galle, 4
Johann Gottfried Galle, 1, 6, 29, 35, 131, 135
John Couch Adams, 5, 29, 33, 131, 136
John Herschel, 30, 31, 34
John Pringle Nichol, 40
Johns Hopkins University, 73
Joseph Fraunhofer, 36, 39
JSTOR, 41
Julian year (astronomy), 2, 3, 117
Julio Ángel Fernández, 53
Jumping-Jupiter scenario, 115
Juno (spacecraft), 147
Jupiter, 8, 26, 30, 60, 88, 95, 98, 108, 109, 112, 145
Jupiter Icy Moons Explorer, 147
Jupiter trojan, 76, 79, 80, 107, 111, 153

Kelvin, 3, 4, 63, 87

Kenneth Edgeworth, 51
Kepler space telescope, 124
Ketu (mythology), 7
Kilogram, 127
Kilometre, 80, 102
Kinetic energy, 87
Kirkwood gap, 56
Kitt Peak National Observatory, 53
Kobe University, 59
Korean language, 7
Kozai mechanism, 113
Kuiper belt, 8, 18, 19, 49, **49**, 76, 86, 90, 93, 107, 112, 126, 148
Kuiper belt object, 26, 96

L3, 80, 81
L4 and L5, 19, 76, 80, 81, 97
Lagrange point, 111
Lagrangian point, 19, 58, 76, 80, 97, 138
Lambertian reflectance, 155
Laomedeia, 119, 122, 128
Laplace plane, 155
Large Synoptic Survey Telescope, 59
Larissa (moon), 10, 22, 118, 120, 127, 132, 146
La Silla Observatory, 131
Late Heavy Bombardment, 106, 107, 109
Lawrence Livermore National Laboratory, 10, 45
Lead, 47
Liberté, égalité, fraternité, 134
Libration, 78
Life, 95
Light-year, 85
List of adjectivals and demonyms of astronomical bodies, 1
List of minor planet discoverers, 1
List of possible dwarf planets, 70
List of Solar System objects by mass, 154
List of Solar System objects by size, 127
List of water deities, 117, 120
Lithium, 86
Local Group, 104
Longitude, 137
Longitude of the ascending node, 2
Longitude of the periapsis, 131
Long-period comet, 53
Luminosity, 101
Lunar node, 7
Lyapunov time, 97

Magnetic dipole moment, 13
Magnetic field, 13
Magnetic moment, 13
Magnetopause, 13
Magnetosphere, 12, 13, 143, 145

Main-belt comet, 92
Main sequence, 87, 97
Makemake, 49, 64, 69, 70, 73
Māori language, 7
Mark R. Showalter, 127
Mars, 59, 88, 96, 100
Mars trojan, 153
Mass, 2, 32, 65, 136
Mass and size distribution, 58
Mass extinction, 103
Mathematical model, 32
Mathematical proof, 109
Matthew J. Holman, 128
Mauna Kea, 54
Mean anomaly, 2
Mean motion, 156
Mean-motion resonance, 93, 109
Media:en-us-Neptune.ogg, 1
Megafauna, 45
Mercury (planet), 88, 91, 101
Merz und Mahler, 39
Metallicity, 86, 105
Meteor Crater, 95
Meteorite, 12, 43, 105
Methane, 3, 4, 9, 10, 43, 45, 49, 63, 69, 123, 133
Methane clathrate, 3
Metis (moon), 98
Metonymy, 9
Micrometre, 134
Milky Way, 103, 106, 132
Millisecond pulsar, 47
Mimas (moon), 122
Mineral, 43
Minor planet, 50
Minor Planet Center, 50, 68, 71, 80
MIT, 53
Moissanite, 47
Molecular cloud, 83, 85
Moment of inertia factor, 2
Mongolian language, 7
Monthly Notices of the Royal Astronomical Society, 53
Moon, 83, 90, 97, 110
Moon–ring systems, 97
Moonlets, 135
Moons, 91
Moons of Jupiter, 69, 120
Moons of Neptune, 2, 4, **117**, 134, 138, 143
Moons of Pluto, 58
Moons of Saturn, 69
Moons of Uranus, 69, 122
Moore, 151
Munich, 39

Nahuatl language, 7

Naiad, 120
Naiad (moon), 21, 118, 120, 127, 131, 134, 146
Naked eye, 23, 30
Nanodiamond, 43
NASA, 15, 26, 143, 147
Natural satellite, 2, 10, 20, 49, 70, 83, 96, 117
Nature (journal), 41, 108
Nebular hypothesis, 83, 84
Neptune, **1**, 29, 30, 43, 49, 51, 55, 56, 60, 68, 76, 78, 80, 88, 92, 108, 109, 112, 117, 118, 131, 132, 143
Neptune (mythology), 3, 7, 120
Neptune Orbiter, 26, 148
Neptune trojan, 19, 49, 58, 76, **76**, 107, 111
Nereid, 120
Nereid (moon), 21, 26, 117, 118, 122, 128, 146
Neso (moon), 117–119, 122, 128
New Frontiers program, 148
New Horizons, 70, 71, 76, 77, 147
New Horizons 2, 26
Newtons law of gravitation, 29, 31
Newtons laws of motion, 31
Nice, 107
Nice 2 model, 115
Nice model, 20, 54, 60, 65, 69, 83, 93, **107**
Nicholas Kollerstrom, 38, 40
Nickel, 88
Night sky, 4, 30
Nitrogen, 4, 123
Nitrogen-14, 44
Nitrogen-15, 44
Noble gas, 43
North America, 45
Notes, 154
Nubian desert, 45
Nuclear fusion, 85
Nucleosynthesis, 43, 86

Observational bias, 59
Observatoire de la Côte dAzur, 107
OB star, 20
Occasional Notes of the Royal Astronomical Society, 41
Occultation, 22, 63, 132
Oceanus, 7
Oldest rock, 105
Olin J. Eggen, 38
Olivine, 48
One dimension, 95
Oort cloud, 49, 51, 52, 76, 93, 96, 103, 106, 107, 114
Opposition (astronomy), 25
Optical depth, 131, 134
Orbit, 5, 31, 49

Orbital eccentricity, 2, 50, 56, 71, 91, 96, 97, 109, 127
Orbital inclination, 2, 71, 91, 117, 124, 127
Orbital node, 156
Orbital period, 2, 53, 127
Orbital resonance, 19, 51, 56, 58, 91, 97, 115, 122, 131, 136
Orbital speed, 2
Order of magnitude, 76, 105, 122
Organic compound, 122, 131, 134
Origins, Dynamics, and Interiors of the Neptunian and Uranian Systems, 148
Orion Nebula, 86, 106
Osculating orbit, 1, 151
Otrera, 79
Outgassing, 64
Oxford, Cambridge and Dublin, 32
Oxford Dictionary of National Biography, 40
Oxygen, 102
Oxygen ion, 10

Panspermia, 92
Pan-STARRS, 59
Paris, 29, 34
Paris inch, 39
Paris Observatory, 31, 131
Parsec, 85
Pascal (unit), 9, 16, 45
Pasiphae group, 125
Patrice Bouchet, 131
Patrick Moore, 26, 40
PBS, 25
Periapsis, 21
Perihelion, 80
Perihelion and aphelion, 1, 2
Periodic comet, 50, 53
Perturbation (astronomy), 3, 5
Phase angle (astronomy), 155
Phobos (moon), 96, 98
Phoebe (moon), 49, 117
Photolysis, 12
Physics, 83
Pierre-Simon Laplace, 83–85
Pioneer 10, 147
Pioneer 11, 147
Plane of the ecliptic, 55
Planet, 3, 6, 29, 35, 83
Planetary core, 10
Planetary embryo, 90, 91
Planetary flyby, 4
Planetary-mass objects, 117
Planetary migration, 20, 56, 60, 83, 89, 91, 92, 107, 109
Planetary nebula, 83, 101, 106
Planetary ring, 4, 22, 98, 131
Planetary science, 83, 107

Planetesimal, 59, 91, 108, 110
Planetesimals, 87
Planets beyond Neptune, 59
Planet X, 26, 52, 145
Plate tectonics, 105
Plutino, 19, 49, 51, 58, 76
Pluto, 19, 49, 51–53, 58, 64, 66, 70, 71, 76, 77, 93, 155
Plutoid, 49, 76
Polar caps, 146
Polar ice cap, 123
Polycrystalline, 47
Polycyclic aromatic hydrocarbon, 45
Polyhedron, 122
Polystyrene, 46
Poseidon, 7, 120
Potential energy, 45
Power (physics), 155
Poynting–Robertson effect, 111
Precession, 124, 156
Prediscovery, 31
President of the Royal Astronomical Society, 37
Presolar grains, 43
Primordial element, 51
Project Gutenberg, 39
Proteus (moon), 10, 21, 22, 117, 119, 120, 128, 146
Protoplanetary disc, 59, 87, 106
Protoplanetary disk, 20, 83, 84, 107
Protostar, 87, 106
Provisional designation in astronomy, 80
Psamathe (moon), 118, 119, 122, 128
PSR J1719-1438 b, 47
PubMed Identifier, 75

Quadrupole, 13
Quaoar, 64
Quasi-satellite, 19, 78

Radiation, 131, 134
Radiometric dating, 105
Red giant, 83, 85, 97, 101, 106
Red giant branch, 101, 106
Reemergence of the nebular hypothesis, 107
Refracting telescope, 39
Refractor, 6
Regolith, 100
Regression analysis, 32
Regular moon, 117
Regular satellite, 112
Resonances, 55
Resonant trans-Neptunian object, 49, 57, 76, 81, 107
Retrograde and prograde motion, 117, 123, 155

Retrograde orbit, 21, 68
Richard Sheepshanks, 36
Right ascension, 3
Ring nebula, 102
Rings of Jupiter, 131, 134
Rings of Neptune, 117, 120, **131**, 143
Rings of Saturn, 22, 99, 131, 134
Rings of Uranus, 118, 131, 132, 134
Robert Hazen, 48
Roche limit, 21, 99
Rock (geology), 49, 123
Roman mythology, 7
Roman numerals, 155
Rotating frame, 78
Rotation, 13
Rotation period, 2
Royal Astronomical Society, 36
Royal Observatory, Greenwich, 6, 33
Royal Society, 29, 36
Russian Academy of Sciences, 7

Sao (moon), 119, 122, 128
Satellites, 69
Saturn, 9, 25, 26, 49, 60, 88, 98, 101, 108, 109, 112, 117
Scale height, 3
Scattered disc, 49, 50, 54, 56, 57, 70, 76, 93, 114
Scientific American, 38
Scientific theory, 51
Scott S. Sheppard, 128
Scott Tremaine, 53
Sears C. Walker, 31
Secular resonance, 110
Sednoid, 49, 57, 76
Semi-major and semi-minor axes, 2
Semi-major axis, 57, 58, 127
Shepherd moon, 120, 131, 135
Shock compression, 45
Shock wave, 86
Short-period comet, 53
Sidereal time, 39
Silicate, 10, 88, 108
Silicon carbide, 43, 46
Small Dark Spot, 14, 15
Small Solar System bodies, 107
Small Solar System body, 49, 79, 83
Solar day, 2
Solar mass, 86, 87
Solar nebula, 51
Solar System, 3, 37, 43, 49, 83, 107, 112, 117, 135
Solar wind, 13, 89
Solid nitrogen, 64, 146
Soot, 146
Space age, 83

169

Space Launch System, 73, 148
Space weathering, 96
Spectroscopy, 62
Spectrum, 62, 122
Sphere, 146
Spheroid, 20, 126
Spiral arms, 104
Spitzer Space Telescope, 45, 124
Star, 35, 43
Starburst galaxy, 104
Stardust (spacecraft), 89
Star formation, 87
Stellar evolution, 87, 97
Stellar nucleosynthesis, 85
Stellar wind, 89, 101
Stratosphere, 11
Streaming instability, 60
Sudan, 45
Sun, 2, 3, 14, 19, 44, 49, 76, 80, 83, 112
Superionic water, 10
Supermassive black hole, 104
Supernova, 43, 47, 86, 101
Supernovae, 104
Supersonic speed, 13
Surface area, 2
Surface energy, 44
Surface gravity, 2, 8, 64
Synchronous orbit, 120
Synchronous rotation, 21
Synonym, 55

Tangaroa, 7
Telescope, 4, 5, 30, 39
Temperature, 3
Template:Distant minor planets sidebar, 49, 76
Template:Human timeline, 95
Template:Nature timeline, 95
Template talk:Distant minor planets sidebar, 49, 76
Terrestrial planet, 88, 109
Terrestrial planets, 88
Tesla (unit), 13
Thai language, 7
Thalassa (moon), 21, 118, 120, 127, 131, 134, 146
Thales Alenia Space, 73
Theory, 84
Theory of relativity, 85
Thermosphere, 11, 12
The Sun and planetary environments, 98
Tholin, 64
Three-body force, 112
Three-body problem, 126
Tidal acceleration, 21
Tidal bulge, 98
Tidal deceleration, 98, 99, 101, 120, 124

Tidal force, 104
Tidally locked, 98, 99
Tidal tails, 104
Tide, 98
Timeline of discovery of Solar System planets and their moons, 127
Timeline of the evolutionary history of life, 94
Timeline of the far future, 84
Titan (moon), 25, 96, 101, 125
Tlaloc, 7
Torus, 56
Trans-Neptunian object, 49, 51, 55, 76, 92, 97, 109, 153, 155
Trans-Neptunian objects, 155
Trident, 3
Triton (moon), 4, 20, 26, 29, 49, 64, 68, 70, 96, 98, 99, 112, 117, 118, 122, 124, 128, 132, 135, 144, 146
Trojan (astronomy), 19, 76, 111
Trojan War, 79
Tropopause, 11, 16
Troposphere, 11, 16
T Tauri star, 87, 106
Tunguska event, 95
Twotino, 19, 58

Ultraviolet, 12
Ulysses (spacecraft), 147
University of California Berkeley, 45
University of California Santa Barbara, 45
University of Hawaii, 54
University of Melbourne, 5, 30
University of Nottingham, 27
University of St. Andrews, 40
Uranus, 3, 5, 8, 16, 29, 31, 32, 43, 46, 52, 58, 60, 88, 92, 108, 109, 112
Urbain Le Verrier, 1, 4–6, 29, 34, 36, 131, 135
Ureilite, 44
U.S. Naval Observatory, 31

Vanth (moon), 97
Venus, 88, 98, 101
Villanova University, 132
Visible spectrum, 12
Volatiles, 8, 9, 49, 108, 123
Volatility (physics), 53
Volcanism, 105
Volume, 2
Vortex, 16
Voyager 1, 25, 147
Voyager 2, 4, 25, 38, 46, 73, 99, 118, 131, 133, 143, 147

Washington, D.C., 48
Water, 49
Water (molecule), 3

Wavelength, 133
White dwarf, 47, 83, 102, 106
Wikipedia:Please clarify, 14
Wikt: evolution, 83
Wilhelm Struve, 31
William Herschel, 7, 29, 31
William Lassell, 21, 29, 117, 118, 128, 131, 132, 135
W. M. Keck Observatory, 23, 138

Yarkovsky effect, 111

CPSIA information can be obtained
at www.ICGtesting.com
Printed in the USA
LVHW111645100922
728083LV00021B/429

9 789352 979561